D0722501

Strange Bedfellows

ALSO BY THE SAME AUTHORS

The Myth of Monogamy

Making Sense of Sex

Strange Bedfellows

The Surprising Connection between Sex, Evolution and Monogamy

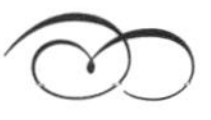

By David P. Barash, Ph.D.
and Judith Eve Lipton, M.D.

BELLEVUE LITERARY PRESS
NEW YORK

First published in the United States in 2009 by
Bellevue Literary Press, New York

FOR INFORMATION ADDRESS:
Bellevue Literary Press
NYU School of Medicine
550 First Avenue
OBV 640
New York, NY 10016

This book was published with the generous support of
Bellevue Literary Press's founding donor the Arnold Simon Family Trust,
the Bernard & Irene Schwartz Foundation, Jan T. and Marica Vilcek,
and the Lucius N. Littauer Foundation.

Cataloging-in-Publication Data is available from the Library of Congress

Barash, David P.
Strange bedfellows : the surprising connection between sex, evolution and monogamy / David P. Barash and Judith Eve Lipton.
p. cm.
Includes bibliographical references and index.
1. Sex. 2. Sex customs. 3. Sexual behavior in animals. 4. Monogamous relationships. I. Lipton, Judith Eve. II. Title.
HQ23.B27 2009 305.84'22—dc22 2009030230

Book design and type formatting by Bernard Schleifer
Manufactured in the United States of America
ISBN 978-1-934137-20-8 hc
FIRST EDITION
1 3 5 7 9 8 6 4 2

To Anne and Nat Barash

Contents

Strange Bedfellows

Chapter 1

THE TRIALS OF TRINCULO

POLITICS, IT IS SAID, makes strange bedfellows. The phrase actually comes, however, from Shakespeare's play *The Tempest*, and was originally somewhat different. A drunken jester named Trinculo has just been shipwrecked on Prospero's magic island. The storm is still raging, and Trinculo seeks shelter under the ragged garments of the resident monster, Caliban. "Misery," says the resigned but desperate Trinculo, "acquaints a man with strange bedfellows. I will here shroud till the dregs of the storm be past." Centuries later, in 1850, one Charles Dudley Warner, editor of the *Hartford Courant*, modified Trinculo's tribulation to "politics makes strange bedfellows." Equating politics with misery, it appears, is nothing new. Our point is that biology, too, creates strange bedfellows, especially when it comes to mating and marriage.

It is endlessly fascinating to learn who ends up in bed with whom, and for how long, and with what secrets. There are always great surprises, scandals, and scenes. The princess is betraying the prince with a dashing horseman, as the prince betrays the princess as well. The heir apparent falls for a "common" woman and gives up his kingdom. The president has a messy sexual relationship with a young intern. One gorgeous movie star abandons his lovely wife and children for another spectacular movie star—of one sex or another. Other celebrities boast of their promiscuity and sexual track records. This is the stuff of *People* magazine and other tabloids, because—let's face it—people want to know who are the strange bedfellows du jour.

It seems that, by contrast, monogamy is boring, not only for each couple but for the gossiping public. Who cares if two people have been faithfully devoted for decades? It is the departures that make the headlines, except for those rare cases in which a mateship persists for 75 years or so. The point of this book, however, is that monogamy is interesting in itself, and that the strange bedfellows who make it permanent may not necessarily do so in a context of misery. In the 1960s, a pair of our friends were accused of living in "unmarried bliss." The question presents itself: Is it possible to be married and blissful, or at least monogamous and satisfied? Is it possible for human beings to be monogamous at all?

Certain situations—typically, demanding ones such as storms or politics—predispose toward outcomes that involve unlikely combinations. The psychology of mating also creates situations of this sort, because if nothing else, it is highly unlikely that the inclinations of any one individual will be identical to those of another, whether or not the two are matrimonially united. Instincts and desire are actually composites of many things, some of them pointing in contradictory directions. Specifically, as we shall see, human biology predisposes *Homo sapiens* of both sexes to seek sexual variety and multiple partners, thus undermining monogamy. But at the same time, the chief argument of this book is that biology does not preclude monogamy; in fact, there are certain evolutionary factors that resonate strongly *with* monogamy, offering shelter from infidelity's storms.

And make no bones about it: Infidelity makes waves. This is not to say that open relationships or other sorts of arrangements don't also cause mayhem, but rather that infidelity (cheating, deception, philandering) nearly always causes substantial trouble for human beings, as for other animals. Deliberate and careful monogamy can provide shelter from these storms of betrayal, but not without taking its own toll.

Monogamy itself can be tempestuous, among animals no less than human beings, and just as biology often threatens to capsize so many matrimonial boats, it also—ironically, but no less genuinely—offers at least the prospect of stability. Either way, everyone is Trinculo, seeking refuge under less than perfect circumstances. And to understand Trinculo, it helps to look hard at Caliban, and perhaps to conclude, as did Prospero in *The Tempest*: "This thing of darkness I acknowledge mine."

A decade ago, in *The Myth of Monogamy: Fidelity and Infidelity in Animals and People*,[1] we made the case that monogamy is extremely rare in the animal world, that it simply isn't "natural"—for animals or for human beings. The response was dramatic: fascination mixed with outrage; vindication for some, apoplexy for others. We did dozens of radio and television interviews, including appearances on the *Today* show, Bill O'Reilly, Hannity and Colmes, a multipage spread in *People* magazine, a "Questions For" segment in *The New York Times Sunday Magazine*. *The Myth of Monogamy* clearly touched a nerve: If monogamy is unnatural, readers wondered, is it hopeless? Could it be bad? Is infidelity inevitable? What does it mean to say that biology dictates or determines, or—more accurately—influences or predisposes, in one direction or another? So what if it does? And what about the paradox of a happily married husband and wife writing that monogamy is in any sense a "myth"?

In *Strange Bedfellows*, we'll try to answer these questions, in the process exploring the multifaceted, seemingly contradictory, and ever-intriguing connections between biology, sex, and monogamy among human beings and other animals. We may also reassure and empower anyone who desires monogamy but is also nervous about his or her prospects. Whereas *The Myth of Monogamy* pointed out the rarity as well as the difficulty of monogamy—a look at the glass half empty—

Strange Bedfellows mostly examines the glass half full: how monogamy works among those animal species that are successful at it, and thus, how biology, despite its predisposition against monogamy, also leaves substantial room for it. We'll also consider how such "natural lessons" can be applied to human beings. And we'll conclude with some practical advice for people, like ourselves, who take biology seriously—and monogamy, too.

Human beings are complicated creatures, enmeshed in language, symbolism, cultural patterns and expectations, social do's and don'ts, a knowledge of the past as well as hopes for the future, all of which have implications for mateships and monogamy. But whatever else they are, people are also perfectly good mammals, biological entities no less than the products of social learning, and as such, our species is "naturally" resistant to monogamy. No matter how much some individuals may desire it, the reality is that most members of *Homo sapiens* have a hard time reconciling themselves to a lifetime with one and only one adult partner.

Nonetheless, it is paradoxically true that biology can also help make monogamy work.

When it comes to romance and matrimony, Freud was wrong: Biology is *not* destiny. Or at least he was wrong in the simplest sense, because genetic programming is not destiny. Indeed, there is no destiny, in part because there is very little "genetic programming," in the sense of rigidly circumscribed limits, especially when it comes to human behavior. Biology is simply the fundament of life, its basic hardware and a lot of important software as well. But "wetware," the complex whole that transpires within the brain, including thinking, feeling, perceiving, and making decisions, is composed of much more than biology at its most reductionistic. To be sure, all wetware activity is ultimately composed of a gazillion interactions of DNA, proteins, and other molecules, and all these molecules are composed, in turn, of atoms, and all of the atoms obey the basic laws of physics. However, it is pointless to try to understand human behavior based on atomic or nuclear physics, and it is almost equally absurd to try to make sense of dating and mating based on DNA alone. Freud might as well have said that physics is destiny—which would have gotten us no farther.

Biology does, however, create conflict between individuals who are each necessarily striving for their own goals, whether they know it or not. Thus, the decisions of each person has consequences that are unique and sometimes troublesome, not only for the individual in question, but also for his or her partner, if partner there be. Sauce for the goose, in short, need not be sauce for the gander. And in this sense, all people are geese.

It is precisely when human beings are misinformed about their biological predispositions that they are most vulnerable to misery. Think about the generations of young persons who didn't know that sex made babies, and therefore had little control over pregnancy. Today most people know at least the rudiments of their sexual biology, and many even understand contraception and how to use it. However, most don't understand the biology of desire, and are as ignorant as our forefathers and -mothers used to be about the basic plumbing of reproduction.

People may well enter into a relationship, for instance, assuming that all temptations are instantly behind them because they have found the love of their lives, only to be blindsided by their all-too-human nature. The "seven year itch" happens (sometimes sooner, sometimes later): Someone cheats, out of anger, temptation, or plain boredom, after which betrayer or betrayed may well conclude—incorrectly—that he or she is "just not cut out for monogamy."

The reality is more complicated, and challenging, because of the biology we all share. For one thing, probably *no one* is "cut out" for monogamy. And yet it *can* be achieved—and even enhanced—not just in spite of our biological legacy but *because* of it.

First, the shocking reality: There is no question about monogamy being natural. It isn't.

There is also good news, however. Some of the best things people do involve going *against* their instincts. And even though monogamy is unnatural, and therefore difficult, people *can* and *do* make it work. It is an everyday miracle.

Books on marriage and relationships invariably ignore evolutionary biology altogether, or they get it wrong, falling into myriad misunderstandings of how evolution works. The most common mistake is to think that "natural selection" means "the survival of the fittest," taken to mean the survival of the physically strongest or most powerful. Another common mistake is to assume that Darwinian selection is by necessity brutal and selfish. The truth is that much behavior is cooperative. Furthermore, "fitness" does not mean cardiovascular strength or bench-pressing prowess, it means success in projecting one's genes—or, from the perspective of a gene, one's self—into the future. This doesn't mean, however, that people are condemned to be nothing more than genetic catapults, fleshy means whereby their DNA makes copies of itself.

Harnessed to conscious human understanding and directed toward particular goals, understanding our biological selves can make a "miracle"—something of transcendence, harmony, and great beauty, nothing literally supernatural, but rather something memorable in which nature, with human help, surpasses its apparent limitations. To the technologically naïve, an airplane is a miracle, at least in part, because a heavy, flying, man-made object is so profoundly "unnatural." But airplanes work, and wonderfully well, in large part because they take gravity into account. Similarly, a would-be monogamist had better consider biology.

This leads to an important issue, one familiar to ethicists but all too often misunderstood by the public: the pros and cons of "natural." It's easy (one is tempted to say "natural") to equate *natural* with *good*. After all, growing numbers of people value—even treasure—natural environments, natural inclinations, and natural remedies, equating "unnatural" with uncomfortable, artificial, and undesirable, even dangerous. Protecting the world's resources and ecosystems—indeed, saving the world itself, not from itself but from *Homo sapiens*—is an important part of cherishing that which is natural.

It is an ethos that applies at the personal level, too. Our own preferred recreational activities, like horseback riding, swimming, and hiking, embed us in nature. We—David Barash and Judith Lipton—have also surrounded ourselves with animals of all sorts, and we try to avoid consuming pesticides, herbicides, and the antibiotics and hormones to which industrial agriculture has become addicted. We were delighted when a natural foods supermarket opened within a mile of our home, and we now patronize it almost exclusively.

Nonetheless, in resisting many things that are unnatural, such as nuclear weapons, global warming, and chemical pollution, while also honoring, respecting, defending, validating, supporting, admiring, and nearly worshiping many things that are natural (sometimes even *because* they are natural), it is all too easy to get carried away, to forget that much in the world of nature is unpleasant, indeed odious and downright despicable. Consider typhoid, cholera, polio, plagues, or AIDS. What can be more natural than viruses or bacteria, composed as they are of proteins, nucleic acids, carbohydrates, and the like? Ogden Nash said it succinctly: "Smallpox is natural, vaccine ain't." To be sure, there are people who object to vaccination, but they would likely object far more to getting smallpox.

We recall returning soaking wet, cold and miserable, more than half hypothermic after a backpacking trip in the gloriously natural Canadian Rockies, during which fog and mist had alternated with rain, hail, and snow (in August, mind you!), and then coming across this bit of wisdom from 19th-century English writer and art critic John Ruskin: "There is no such thing as bad weather, only different kinds of good weather"—whereupon we concluded that Mr. Ruskin hadn't spent much time in the mountains. Similarly, we suspect that those well-intentioned people who so admire "natural" raw milk have never experienced the ravages of *Campylobacter*, pathogenic *E. coli*, or bovine tuberculosis, each spread by unpasteurized milk.

Even in sports, with its cult of the "natural" athlete, devotees strive to move beyond the natural to what is beautiful, elegant, or impressive, fully recognizing that it takes work and practice. That's why there's spring training, exhibition games, coaches, trainers, and interminable "practice." Dressage (a classical form of horsemanship) seeks to help a horse move with a mounted rider as beautifully as it would solo, in nature. To do this takes at least a decade of effort, pushing horse and rider to work hard and in unnatural ways in order ultimately to achieve harmony and beauty. It is natural for horses to stand around in fields, eating and pooping and swatting flies; it is not natural for a horse to dance to music. So the training of the horse brings out its natural beauty, but only after enormous amounts of unnatural labor. Similarly, it is not natural for people to climb high mountains. It is difficult and stressful, requiring physical conditioning, skill, and technical equipment. But for many, getting to the top of a mountain is a "peak experience" indeed, worth the time, trouble, and risk.

In short, natural is often good, but not always. Sometimes (more often than many "naturalists" would think) unnatural is better. It may be natural to punch someone in the nose if he has angered you, for people to get sick, or for a child to resist toilet-training. And of course, bacterial infections, lousy weather, and awkward behavioral inclinations aren't the only troublesome entities out there in the natural world. Don't forget about hurricanes, tsunamis, earthquakes, droughts, the devastation wrought by volcanoes, lightning storms, sandstorms, and blizzards. But for the purposes of this book, and for much of the philosophical debate that has swirled around the subject, the supposed desirability of "the natural" refers to behavioral inclinations.

In his book *A Treatise on Human Nature*, published in 1739/1740, Scottish philosopher David Hume presented, and criticized, what has come to be known as the "is-ought problem," the notion that we can derive what *ought* to be from an examination of what *is*. Is there any way, Hume asked, that we can legitimately connect how the world "is" (which by extension includes our own behavioral inclinations) with how it "ought" to be (including how we ought to behave)? At least one respected modern philosopher feels that simply by raising the question, Hume so conclusively severed "is" from "ought" that he called this distinction—between the *descriptive* and the *prescriptive*, or between *facts* and *values*—"Hume's Guillotine."[2] Hume's insight, that it is fallacious to derive "ought" from "is," has come to be known as the "naturalistic fallacy," a term that was coined by English philosopher G. E. Moore, in his 1903 book *Principia Ethica*.

In 1710, three decades before Hume sliced into the is-ought problem, German philosopher and mathematician Gottfried Leibniz had struggled with the problem of "theodicy," the theological effort of reconciling the existence of evil and suffering in a world supposedly governed by a god that is both all-powerful and wholly benevolent. It was—and still is—a tall order. Leibniz concluded, to make a very long story misleadingly short, that since god is necessarily good (by Judeo-Christian definition), as well as omnipotent, and since the deity evidently chose to make the world as it is, in view of all the possible ways that it might have been, then this must be "the best of all possible worlds" (*le meilleur des mondes possibles*). This famous phrase has proven easy to satirize, most notably by Voltaire, in his novel *Candide*, the picaresque adventures of Dr. Pangloss (Voltaire's Leibniz caricature) and his student Candide, who experience no end of terrible events—always interpreting them through a cheerful, positive lens.

Voltaire was especially outraged by a particularly devastating natural disaster, the Lisbon earthquake of 1755, which is estimated to have killed tens of thousands of innocent people. But he also wasn't shy about depicting the outrageously cruel but equally "natural" behavior of murderers, rapists, and torturers. It's a theme that continued to resonate: In the 19th century, John Stuart Mill argued in his famous essay "Nature" that "nature cannot be a proper model for us to imitate. Either it is right that we should kill because nature kills; torture because nature tortures; ruin and devastate because nature does the like; or we ought not to consider what nature does, but what it is good to do."

Modern evolutionary biology makes it clear that "nature," acting through natural selection, whispers in our ears—cajoling, seducing, imploring, sometimes even threatening or demanding—and undoubtedly inclining us in one direction or another. These inclinations, in turn, are derived from a remarkably simple process: the automatic reward that comes from biological success. If a given behavior leads to greater eventual reproductive success on the part of the "behaver" (more crucially, heightened success for any genes that predispose toward the act in question), then selection will promote those genes, and thus, the behavior. It will seem—and be—natural.

Natural selection has a very efficient way of getting animals and people to do things that are "good" for the organism—or at least, things that contribute to the success of those genes that generate a propensity for doing those things: call it pleasure. Living things find it pleasurable to eat when hungry, drink when thirsty, sleep when tired, obtain sexual satisfaction when aroused, and so forth. The evolution of genes for, say, self-nourishment would not be well served if those genes induced their bodies to refrain from eating. And so, eating when hungry or drinking when thirsty feels "good" because feeling good is what gets animals and people to do those things, and organisms that didn't make the "do such-and-such/feel good" connection would have left fewer descendants than those for whom pleasure basically tracked reproductive success, what evolutionists call "fitness." But whether, in Mill's terms, such things are necessarily "good to do" in the sense of ethics and morality is another matter entirely.

Gravity exists, quite naturally. But few people would derive ethical guidance from the natural world of physics, which would mean no more standing upright, since this goes against such a universal moral precept. Should we refrain from cleaning the house, since the Second Law of Thermodynamics—another fundamental, natural law—dictates that disorder necessarily increases within any closed system, and therefore entropy is good and struggling against it is wrong? Is it unethical to exceed the speed of light, or simply impossible? Similar absurdities arise if one attempts to "naturalize" ethics from chemistry, geology, astronomy, mathematics, and so forth. When it comes to biology, however, many people seem to feel otherwise.

After all, isn't there something good—maybe even magnificent—about brilliant autumn foliage, the song of a nightingale, the majesty of a bull elephant? If nothing else, they bring pleasure, even delight, to people. And isn't it downright *good* for a mother robin to feed her nestlings? Doing so is certainly good for the baby robins and, thus, for the evolutionary success of the adults. Leaving aside, for the moment, the less-than-salubrious effect this has on those worms whose lives are thereby cut short, it is easy to assume that the working of biological nature—as distinct, perhaps, from physical nature, or chemical nature, or geological nature—is not only admirable at the level of observing human intellects but also ethically instructive.

But wait! A dispassionate look at the natural world seems to lead, if anything, to confirmation that the naturalistic fallacy is indeed fallacious. Not that it denies basic logic to say that if something is biologically natural it must be good; it isn't equivalent to concluding, say, that if Socrates is a man and all men are mortal, then Socrates is *not* mortal. Rather, it simply does not follow that biological nature is necessarily good, in the sense of giving us insight into morality or ethics. To an extent that should trouble any "natural ethicist," the living world is a zero-sum game, in which benefit for one organism comes at the expense of others, and no sign of overarching ethical restraint, no independent claim to goodness, can be discerned.

Annie Dillard's marvelous meditation on nature, *Pilgrim at Tinker Creek*, contains an unforgettable account of her encounter with a "very small frog with wide, dull eyes." As Dillard describes it:

> just as I looked at him, he slowly crumpled and began to sag. The spirit vanished from his eyes as if snuffed. His skin emptied and drooped; his very skull seemed to collapse and settle like a kicked tent. He was shrinking before my eyes like a deflating football. I watched the taut, glistening skin on his shoulders ruck, and rumple, and fall. Soon, part of his skin, formless as a pricked balloon, lay in floating folds like bright scum on top of the water: it was a monstrous and terrifying thing. I gaped bewildered, appalled. An oval shadow hung in the water behind the drained frog; then the shadow glided away.

This "shadow," which had just killed the frog—so smoothly, mercilessly, and naturally, before gliding away—was an enormous, heavy-bodied brown creature that (to quote Dillard, once again):

> eats insects, tadpoles, fish, and frogs. Its grasping forelegs are mighty and hooked inward. It seizes a victim with these legs, hugs it tight and paralyzes it with enzymes injected during a vicious bite. . . . Through the puncture shoot the poisons that dissolve the victim's body, reduced to a juice. This event is quite common in warm fresh water. The frog I saw was being sucked by a giant water bug. . . . I stood up and brushed the knees of my pants. I couldn't catch my breath.

Dillard quickly goes on, achieving a more objective and scientific detachment:

> Of course, many carnivorous animals devour their prey alive. The usual method seems to be to subdue the victim by downing or grasping it so it can't flee, then eating it whole or in a series of bloody bites. Frogs eat everything whole, stuffing prey into their mouths with their thumbs. People have seen frogs with their wide jaws so full of live dragonflies they couldn't close them. Ants don't even have to catch their prey: in the spring they swarm over newly hatched, featherless birds in the nest and eat them tiny bite by bite.

She notes, with understatement, that "it's rough out there, and chancy," that "every live thing is a survivor on a kind of extended emergency bivouac," and that "cruelty is a mystery," along with "the waste of pain." And finally, Dillard concludes that we "must somehow take a wider view, look at the whole landscape, really see it, and describe what's going on here. Then we can at least wail the right question into the swaddling band of darkness, or, if it comes to that, choir the proper praise."

In Stephen Sondheim's dark musical, *Sweeney Todd*, we learn that the story of the world boils down to who eats and who gets eaten. Biology makes strange bedfellows: pleasure and pain, suffering and delight, eater and eaten, life and death. To this, one might be tempted to add good and bad, although it would be more accurate to conclude *neither* good *nor* bad. Like physics, chemistry, geology, or mathematics, biology simply *is*.

Our point is that when it comes to evaluating monogamy's goodness (or badness), it really doesn't matter whether it is natural or *un*. At the same time, monogamy's naturalness or unnaturalness matters greatly when it comes to efforts at understanding how it sometimes works and sometimes fails, why monogamy exists in some cases and not in others, and what, if anything, can be learned from biology that might expand, contract, or otherwise modify how people engage in this, the most intimate relationship of their lives.

Let's grant, for starters, that monogamy isn't natural, and also that the question is irrelevant to whether or not it is good. Doesn't monogamy's unnaturalness at least bear upon whether it is feasible? After all, many things that people cannot accomplish are beyond their reach simply because we as a species have certain biological limitations. We cannot fly like birds, or hear ultra-high frequencies in the manner of bats or dogs. We cannot live in a vacuum, or without water, or even on bread alone. (We're thinking of nutritional sustenance here, not spiritual.) All these and many other limitations are imposed by our biology, and even though it may be exaltingly romantic to exhort human beings to rise above their organic limitations—for example, to be more than you can be, literally reach for the stars, or hold the world in your hands—it simply can't be done.

Such extreme cases, however, are misleading. *Of course* anything that is sufficiently unnatural is beyond human capability; this is true by definition. But paradoxically, human beings are endowed by evolution with the capacity to do things that are extraordinarily independent of biology, things that do not precisely transcend our organic nature (since, once again, this is definitionally impossible), but nonetheless thumb our collective noses at biological promptings. In this regard, *Homo sapiens* is probably unique in the animal world.

Keep in mind that many things are "unnatural," but do-able, and often desirable. They simply take time, energy, and effort. Playing a musical instrument, for example, isn't natural. It takes practice, especially if it is to be done well. If protracted monogamy is rare, so are good musicians. But both can be beautiful, as well as achievable. Moreover, here is some additional good news: Unlike virtuoso musicianship, the ability to be happily and successfully monogamous does not require extraordinary talent or perfect pitch—it just takes knowledge and determination.

One important difference between musical and sexual preferences, however, is that whereas there is little reason to be in the closet about a private fascination with waltzes or swing dancing, people commonly pretend to value monogamy, while often secretly cheating or discarding relationships. "Ozzie and Harriet" notions of monogamy are so much a part of American social expectations that (even after the never-ending and well-publicized affairs of Hollywood stars and political leaders) the public still sees monogamy without the hard labor, departures, mistakes, reconciliations, and bids for forgiveness and repair that are prerequisites for its success. Worse, many expect it will be easy once wedding vows are exchanged. Not so.

Given this widespread misconception, many people despair when they find themselves, or their partner, interested (even fleetingly) in someone else. The point is that, technically speaking, *no one is cut out for monogamy*, but at the same time, nearly everyone with a functioning frontal lobe is capable of it.

Part of the problem is that monogamy is so highly esteemed and so politically correct—at least in the United States and the West—that many people find it shocking when someone pulls the cover off the façade to point out that genuine monogamy is the exception, not the rule. The Judeo-Christian tradition advertises and typically prescribes—even demands—"till death do us part," and yet, such expectations do not reflect what really goes on in people's homes and bedrooms. Most Americans, for example, begin to date in their teens, with or without sex. Relationships break up, new ones form. Although the dating may become more long term, healthy people nonetheless still switch partners with relative ease until they find "the one." Marriages are then made, frequently with great fanfare, followed (more quietly) by divorces about 50 percent of the time.

Some—indeed, the majority—end up practicing "serial monogamy," not uncommonly sprinkled with adultery, because staying with the same human being for decades can be frustrating, dull, and just downright difficult, and not just when monogamy = monotony. "We sleep in separate rooms," quipped Rodney Dangerfield. "We have dinner apart; we take separate vacations. We're doing everything we can to keep our marriage together."

The monogamy challenge derives largely from human biology, but happily, so does the potential benefit, so that monogamous relationships can be gratifying and even beautiful, with payoffs justifying the hard work.

Is this book, then, a work of social—if not biological—advocacy? Why write a book about the "miracle" of monogamy if we do not, in some sense, recommend it? The reality is that we do like monogamy—for ourselves—and we work at it. But at the same time, we do not necessarily condemn its various alternatives; rather, we are aware that monogamy works best for us. You may feel the same . . . or differently. But either way, and whatever romantic/sexual path you decide to follow, we recommend doing so with the fullest possible knowledge of what your deepest self, your biology, has to offer . . . and to threaten.

"A man's reach should exceed his grasp," wrote Robert Browning, "or what's a heaven for?" Although many reach for monogamy only to fall short, it isn't beyond anyone's grasp. Trinculo doubtless found it stressful to shelter under Caliban's rags, but he did it, and, in the process, he survived the Tempest.

WHAT COMES NATURALLY: ANIMALS

THE PREVAILING JUDEO-CHRISTIAN CULTURAL CODE—ethical, legal, religious—calls for monogamy. And yet, the prevailing biological code—the one inscribed, albeit with complexity and nuance, in human DNA—does not agree. The heirs of Western social traditions, including those with ancient ethnic ties to Judaism and Christianity, inherit two very different sets of instructions for when to mate and with whom: cultural norms that typically expect (and mostly try to enforce) sexual fidelity, cohabiting uneasily with biological whisperings that favor *in*fidelity. Infants may revel in their infancy; adults struggle with their adultery. People from cultures that favor polygyny (one man, several women) or polyandry (one woman, several men) are no less conflicted. Males invariably want more females than they can afford, or they want desirable females who are already mated to somebody else, and females want sex with men other than their husbands, even at great personal risk, if the potential gain is sufficient.

Moreover, departures from monogamy are not simply cases of anomalous, aberrant behavior. For most of the animal world, the norm calls for multiple sexual partners, and indeed, biologists have long understood that monogamy is rare among animals. Out of about 4,000 mammal species, only a handful have ever been called monogamous. The tiny list includes beavers, gibbons, certain foxes, otters, and bats, a few hoofed mammals, some primates—notably the tamarins and marmosets of the tropical New World—a couple of rodents, and some exotic species, about which more later. By contrast, birds have long been the poster children of monogamous fidelity. An oft-cited "fact," first compiled by the great ornithologist David Lack, is that 92 percent of bird species are monogamous.[3] Picture a pair of eagles or geese engaging in prolonged courtship, then collaborating in building a nest, and finally, devotedly, taking turns incubating the eggs and provisioning the young.

This notion penetrated into popular culture, along with the unspoken lesson: "If they do it, so can you." Thus, in the movie *Heartburn*, a barely fictionalized account by Nora Ephron of her marriage to the philandering Carl Bernstein, the lead character complains to her father, who responds, "You want monogamy? Marry a swan!" But now we know that even swans aren't monogamous: In the Australian black swan, for example, one in six cygnets are fathered by someone other than the mother's "official" partner.[4]

Actually, the myth of monogamy didn't disappear overnight. Researchers began detecting the telltale hiss of its deflation several decades ago. During the 1970s, for example, a now-famous study reported on attempts to use vasectomies to achieve nonlethal population control among blackbirds. To their surprise, the researchers discovered that female blackbirds that were mated to vasectomized males were nonetheless laying eggs that hatched![5] Evidently there was some hanky-panky going on in the blackbird world.

And not just there.

Field biologists began to notice that in species after species, even the most happily "married" couples weren't always sexually faithful, leading the scientists to question—tentatively at first, then with increasing confidence—whether social monogamy (one-to-one male-female bonding and cooperation) and sexual monogamy (one-to-one sexual exclusivity) are synonymous. Then, in the 1990s, came DNA fingerprinting and an avalanche of discovery: The same techniques that can identify paternity or prove innocence (or guilt) in rape cases could also demonstrate that a child (or chick) was (or wasn't) fathered by a particular male. Time and again, it was revealed that 10, 20, even sometimes 30 or 40 percent of nestlings were not genetically related to the evident social father; in short, the mother's "husband" wasn't necessarily the genetic father of all her offspring. At the same time, the social mother is guaranteed to be the biological mother, thereby legitimating the old adage: "Mommy's babies, Daddy's maybes."

Nor are mammals exempt. Gibbons—those long-armed, brachiating acrobats of South Asian rainforests—were long thought to be lifetime monogamists. No longer.[6] Ditto for essentially every species that has been investigated with any thoroughness, with exactly one exception: a species of flatworm (*Diplozöon paradoxum*) that lives in the gills of certain freshwater fish, in which male and female literally fuse together at adolescence, remaining sexually faithful ever after.[7]

The question arises, "Why?" For evolutionary biologists, it is mostly half a question: Why do socially mated *females*—of whatever species—engage in the equivalent of "infidelity" or "affairs," euphemistically called "extra-pair copulations" (EPCs)? As to the sexual proclivity of most males, the "why" question has never been in much doubt. Males make sperm, which are extraordinarily small, produced in amazingly large numbers, and typically impose essentially no biologically mandated follow-through in order for reproduction to succeed. As a result, the optimal tactic for males of most species is to be easily stimulated, not terribly discriminating in regard to sexual partners, and generally willing—indeed, eager—to fertilize as many eggs as possible. The more sex, and especially the greater the number of different partners, the greater the number of likely offspring. Hence, natural selection will have favored a male penchant for sexual variety, whether or not the males in question are already mated.

The story is still told in New Zealand about an Episcopal bishop who visited a Maori village, where he was entertained by the local inhabitants. After the feasting, and as everyone was about to retire for the night, the village headman—wanting to show sincere respect for his high-ranking guest—called out: "A woman for the bishop!" Seeing a scowl on the prelate's face, the headman called out, even more loudly: "*Two* women for the bishop!"

Not that evolution has necessarily generated a male preference for multiple partners simultaneously, although to be sure, that prospect is often more than a bit alluring, at least for *Homo sapiens*. Rather, natural selection has almost certainly endowed (or oppressed) males of many species with a deeper—and more biologically understandable—inclination: for multiple sequential partners and, nearly always, an enhancement of sexual enthusiasm when and if such an opportunity presents itself. It has long been known that even when socially disapproved, and/or to the great consternation of those involved, the male libido is aroused by the possibility of sex with someone new. Thus, Byron wondered:

> How the devil is it that fresh features
> Have such a charm for us poor human creatures?

And of course, it isn't unique to human beings. Animals perceive a similar "charm," as evidenced in this libidinous tale of biological perspicacity, involving—of all people—President Calvin Coolidge.

The story goes that President and Mrs. Coolidge were each, independently, touring a model farm. When Mr. Coolidge reached the poultry yard, his guide informed him that "Mrs. Coolidge wanted me to point out that our rooster mates many times every day." The President immediately inquired: "Always with the same hen?" "Oh, no, sir!" replied the farmer, whereupon Coolidge said, "Please point *that* out to Mrs. Coolidge!"

This little vignette might be apocryphal, but its underlying biological message is altogether real. Indeed, evolutionists speak offhandedly of the "Coolidge effect" Present a rooster with just one hen, and his frequency of copulations rapidly subsides. Introduce a new one, and it increases—at least for a time. Ditto for stallions, bulls—in fact, nearly every species in which the phenomenon has been examined. Give a ram a new ewe, and the likelihood is that, sexually, it will produce a new him.

It matters relatively little to the males in question whether they are already mated. To be sure, some costs may be incurred if the extra-pair female partner is also already mated, should the cuckolded male find out and respond aggressively. There may also be costs imposed by a philandering male's current female mate: aggression, the withholding of future sexual favors, and so forth. Nonetheless, biologists are agreed that, on balance, the likely payoff for male infidelity will be substantial, because among species in which males offer some parental care, a successful philanderer not only gets the potential reward of having his genes replicated, he is also likely to parasitize the assistance of one or more males: the ones he has cuckolded. *They* get stuck rearing *his* offspring: a good genetic deal for him, a lousy one for them. Seen from the strictly calculating, results-driven, gene-focused perspective of natural selection, genetic factors that influence males to fertilize as many females as possible will be favored. More copies of such genetic tendencies will accumulate in future generations, compared with the accrued output of males whose one-to-one sexual fidelity results in comparatively fewer offspring. As a result, natural selection is expected to result in males with a wandering eye, an easily aroused libido, and whose commitment to monogamy is only lukewarm at best.

When it comes to females, on the other hand, the benefit side of the infidelity ledger is more obscure. After all, eggs are fewer and more costly than sperm. No surprise here: They are defined as gametes that are relatively large (compared to sperm) and produced in comparatively small numbers. Moreover, compared to sperm—99.99999 percent of which are

ferent males = a greater variety of sperm = more genetic variety among one's offspring = higher long-term reproductive success if the environment changes in the future. Which, of course, it always does. One problem, however, is that there is immense genetic variety among the hundreds of millions of sperm produced by just one male, and it isn't at all clear that the benefit of yet more gene diversity exceeds the potential cost of getting it.

For certain birds, the benefit may be immediate, such as obtaining food from the territory of one's lover, or even being provisioned directly by him. Among ospreys, for example—also known as "fish hawks"—females whose mates don't bring home the salmon are likely to mate on the sly with other males, who bring them food in return. In many cases, the payoff appears to be more indirect, via genetic benefits to the "out-of-wedlock" offspring: By mating with males who are especially hearty, and/or who possess secondary sexual traits that are particularly appealing to other females, would-be mothers can increase the hardihood, as well as the eventual sexual attractiveness, of their offspring, thereby enhancing their own long-term reproductive success. Among barn swallows, for example, a deeply forked tail is a sexually desirable male trait. Females paired to males whose tails are not especially impressive in this regard have been found to be prone to mate on the sly with those fortunate neighboring males whose forked tails were experimentally enhanced.

Male barn swallow, showing deeply forked tail.

Photo by Terry Sohl, sdakotabirds.com

A not-uncommon female strategy, therefore, appears to be as follows. Even in supposedly monogamous species, females may be selected to mate extracurricularly (and on the sly) with males who are especially healthy, wealthy, and, if not wise, at least sexy, thereby gaining additional resources as well as the payoff of producing healthier, sexier offspring, while at the same time remaining "socially monogamous" and thereby also retaining a partner when it comes to parenting.

Anthropologist Sarah Hrdy has suggested an additional evolutionary benefit to female philandering: It is at least possible that among primates in particular, females solicit EPCs in order to buy a kind of tolerance—not from their identified mates (as noted, the exact opposite reaction is expected, and found) but on the part of their extra-pair Lotharios. Males of many species—chimps, langurs, certain macaque monkeys—are often infanticidal toward offspring they have not fathered. By copulating with "outside" males, females could be bribing them into abstaining from such violence toward offspring that might be their own.

Could these factors (some, all, or others not yet identified) be operating among human beings? Yes, indeed.

Think of what makes someone attractive: looks, personality, money/status/power. If a person is "cute," "good-looking," or "sexy," isn't that another way of saying that he or she offers the prospect of good genes, even if neither party is literally planning to make babies with the other? (By the same token, people find good food attractive, typically without knowing anything about the biochemistry of cell metabolism.) Black-caped chickadees usually mate exclusively with their social partners, making them at least somewhat monogamous, most of the time. The exception? Those rare males who are especially dominant over their neighbors, and who in turn are likely to be chosen for sexual trysts by their neighbors' "wives."

Chapter 3

WHAT COMES NATURALLY: PEOPLE

JUST BECAUSE SOMETHING OCCURS in animals doesn't necessarily mean that it is true of human beings. Some fish navigate by detecting changes in the electric fields surrounding them; people can't. Termites digest cellulose; we can't. But if electric orientation or cellulose digestion were nearly universal in the living world, we could be forgiven for expecting them to apply in the human case, too. We would, in fact, be hard-pressed to explain why it wouldn't be so. After all, the most fundamental take-home message from evolution by natural selection is of *continuity*, the deep, simple fact that all living things are connected, not only in the straightforward sense of sharing common ancestry if we go back far enough, but also in that organisms—all of them—are subjected to the same basic principles of evolution by natural selection.

We are all animals at the core, especially when it comes to the fundamental ways in which evolution has acted upon our bodies and—no less—our behavior.

Not surprisingly, therefore, monogamy has always been a tough slog for people, too. Ibsen's Hedda Gabler, in the play named for her, expresses a species-wide discomfort when, complaining about her marriage, she notes that "what I found most intolerable of all . . . was being everlastingly in the company of one and the same person."

Nineteenth-century theatergoers who were shocked—*shocked!*—at this sentiment may have taken comfort in the fact that Ms. Gabler is one of the most unpleasant characters in all of literature. Nonetheless, Hedda's heresy cannot simply be attributed to a disagreeable disposition; the reality is that a Martian zoologist visiting planet Earth would have no doubt that *Homo sapiens* is not readily monogamous: We sport all the hallmarks of a species that, although capable of forming lasting bonds with the opposite sex, rarely does so.

When it comes to lasting social connections among adults, gorillas have only male-female bonds; chimpanzees, only male-male; bonobos, only female-female. And humans? All three, and sometimes none.

Men are males (sperm-makers), women are females (egg-makers). They—men and women—engage in sex. They find certain behaviors attractive (including, but not limited to, sex itself) and others deeply upsetting (such as if their partners have sex with someone else). They form pair bonds. They depart from these bonds, not always, but more often than most would like. Then they sometimes return. They reproduce. They care for their children. They care for each other, at least sometimes. They grow old, either together or alone.

Biologist Julian Huxley, one of the notable polymaths of 20th-century science—who authored an immensely influential, magisterial volume titled *Evolution, The Modern Synthesis**—needed no convincing as to the power of evolution to shape living things. In fact, much of our understanding of the phenomenon of sexual selection derives from his pioneering research. At the same time, Huxley warned against what he called "nothing butism," the misleadingly simplifying assumption that just because *Homo sapiens* are animals, they are nothing but animals. His point was well taken, and still is, but it also deserves to be inverted: The fact that human beings are strongly influenced by cultural traditions, social expectations, and the huge panoply of learned experience does not mean that they are nothing

* The title of Huxley's book inspired another, equally consequential tome: Edward O. Wilson's *Sociobiology: The New Synthesis* (1975, Harvard University Press).

but the products of those traditions, expectations and experience. Whatever else they are, and regardless of their aspirations and inspirations, people are also animals. And no small part of that animal nature has to do with sex, which means all that maleness, femaleness, and the underlying, albeit often unconscious, reproductive considerations that pertain to other mammals.

In short, people may yearn for monogamy—or claim that they do—but to a disconcerting degree, human biology says otherwise.

Even in its definition, monogamy isn't simple. It comes in different flavors. In its most straightforward (although, paradoxically, most demanding) form, monogamy is sexually exclusive and lifelong, the way geese and swans were thought to mate before biologists learned better. It can also be temporary or serial: The movie *March of the Penguins*, lauded by many as a testament to monogamy, described the extraordinary domestic one-to-one devotion of Emperor penguins, but without mentioning that these animals form new mateships—nearly always with *different* partners—the following year!

Serial monogamy of this sort is also the most frequent human version. Sexual enchantments, short but sweet, are common, especially among adolescents and young adults. Each summer vacation, year abroad, new school, or new semester can bring a thrilling romance, intense and evanescent. Although people do not typically re-establish new domestic partnerships annually, they often live "monogamously" with a single partner until one or the other dies,

Emperor penguin parents and chick.

PHOTO BY ARCTICPHOTO

leaves, or is abandoned or divorced, whereupon a new "monogamous" relationship may be established—often after a period of "dating," in which other potential partners are explored. For literally centuries, monogamy was also interrupted by death, most commonly women dying in childbirth, after which the surviving husband would remarry. Young men died, too, often in wars or conflicts with other men, leaving behind young widows. Thus, women were also serial monogamists, if they could find new husbands despite being no longer young or virginal.

It was generally easier for widowers—especially wealthy ones—to find wives than for widows to find husbands, because for the most part, there weren't as many wealthy women. Also, few men would raise another's children, while it was not uncommon for men to marry the babysitter or housekeeper who had been taking care of his children before his wife died. Even when a presumably monogamous marriage continued over time, strict sexual fidelity was never a universal expectation. "Indiscretions" were often discretely ignored.

But at least serial monogamy offers the pretense of a lifelong, devoted partnership—until it ends. (Friedrich Max Müller, a renowned 19th-century German scholar of Hindu religious thought, coined the term "henotheism," referring to belief in one god . . . at a time. A suitable neologism for serial monogamy might accordingly be "henogamy.") The reality, in any event, is that lifelong monogamy is exceedingly rare among *Homo sapiens*. How many readers of this book have had only one boyfriend or girlfriend in the course of their lives?

Perhaps this apparent predisposition against marrying one's first "serious" partner is itself an unconscious bit of sociosexual wisdom: It is usually a bad idea to mate permanently with one's first lover, if only because youth and inexperience don't promote great decision making. Better to experiment, learn who you are and what you want, who is out there and available and for what, than to make a foolish and premature commitment.

Whereas social conservatives like to point to a supposed growing threat to "family values," they don't have the slightest idea how great that threat really is, or where it comes from. The monogamous family is very definitely under siege, but not by government, declining moral fiber, or some vast underground homosexual agenda, but from the evolutionary biology of living things, which definitely includes *Homo sapiens*. Not that the outcome is gloomy or foreordained. As we'll see, to be besieged is not necessarily to be defeated and overrun; there are countervailing biological forces already enlisted in defense of monogamy, and additional ones that can be rallied. But first, wannabe monogamists need to know their enemy: themselves.

Let's begin with some sobering, biologically based reality. Human beings probably never occupied an Edenic Adam-and-Eve paradise of one-to-one fidelity. They—we—are naturally polygynous, and, paradoxical as it might seem, naturally polyandrous as well, although the polyandry in particular has long been secretive. (Polygynous means one female/many males, while polygamy technically includes both polygyny and polyandry, that is, any arrangement in which a man has more than one wife, or a woman has more than one husband. Polyamory is a very new term, referring to multiple affectionate sexual partners without marriage, including, in most cases, open acknowledgment of these relationships.) Another way of saying the above, therefore, is that people are naturally polygamous.

The evidence is as follows. First, men are significantly larger than women, a pattern consistently found among polygynous species. From deer to seals to primates, the harem-keeping sex is the larger one, because competition among harem-keepers rewards those who are larger and brawnier and more successful in head-to-head competition.

Second, around the world, men are more violent than women, which, as with greater size, reflects their struggle to achieve mating success in a harem-based system in which the harem-keeper achieves his station only after besting his competitors. No one should be surprised, therefore, that harem-keepers are likely to be drawn from the pool of individuals that are not only somewhat larger than the unsuccessful bachelors, but also more aggressive. Since these individuals—larger as well as more violence-prone—were the ones who left more descendants, the result has been selection for harem-keepers who, on average, are larger and more aggressive than the harem-kept. At this point, some readers will point out that they know a woman who is six feet tall, and/or a man who is five foot four. No doubt. There are also some women who are stronger and more violent than some men, and sometimes it snows in April. But generalizations, by definition, do not purport to cover all cases; that's why they're generalizations, not incontrovertible, never-fail rules.

Third generalization: Girls become sexually mature earlier than do boys. This is yet another telltale sign of polygyny, since the intense competition to which harem-keepers are subjected conveys an evolutionary payoff for the "keeping" sex to delay maturation until individuals are large, strong, and possibly canny enough to have some chance of success. It is an interesting and counterintuitive observation, if you think about it. After all, women—and female mammals generally—are the ones that get pregnant, must nourish a developing fetus from their own bloodstream, undergo childbirth, and then provide additional investment in the form of milk, which turns out to be the most metabolically demanding requirement of all. In view of all this, wouldn't it make sense for females to hold off on becoming mothers until they are especially large and strong, at least compared to males, whose biology merely mandates that they produce a teaspoon-full of tiny sperm? Moreover, after such labor they can simply walk away, leaving all the really demanding work to the females.

But this ignores the reality of polygyny, which requires that successful fathers are those that compete successfully with other would-be fathers. And the greater the "degree of polygyny"—the greater the average harem size, or, more technically, the ratio of females to males in the average breeding unit—the greater the male-male competition. Not surprisingly, therefore, among harbor seals—in which the breeding unit is typically one male to one female—the sexes become mature at about the same age. And bull elephant seals—among whom a successful harem master may breed with more than thirty females, typically don't become sexually mature until around eight years of age. Females? About three. The pattern persists among nearly all mammals: The small, mostly monogamous forest-dwelling deer of India undergo sexual maturity at about the same age, whereas among North American elk, with their large harems, bulls take significantly longer to breed than do cows.

This pattern of "sexual bimaturism" is a reliable indicator of breeding system, with the later-maturing sex being the one that competes more vigorously among themselves for access to numerous members of the quicker-maturing sex. The cost of engaging in such competition when too young and too small is evidently so great that the harem-keeping sex is selected to defer entering the reproductive fray. And, of course, girls mature several years earlier than boys, leading to that peculiar early teenage phenomenon in which boys are almost comically conspicuous "lagging indicators."

For a fourth signifier of human polygyny, consider that prior to the cultural homogenization that came with Western colonialism, more than three-quarters of all human societies were polygynous. Monogamy was the preferred norm in fewer than one-quarter of human societies.[8]

There is little doubt that when it comes to their "nature," human beings are mildly polygynous, certainly not monogamous.

But it's one thing to conclude that our biology favors a degree of polygyny and quite another to think that most people, throughout most of time, were either harem-keepers or harem members. The likelihood, in fact, is that only a few well-positioned males succeeded in polygyny, just as only a small proportion of females were chosen (or coerced). The great majority of people—of both sexes—doubtless practiced serial monogamy, or at least its social variety, for lack of better options in a difficult environment. Females were bundled off in marriage by their families as soon as they passed menarche, and they often died in childbirth. Endless pregnancies and lactation would not have left much time for flirtation, and it has never been easy to be a single mother; so, better to pair with someone, even if he turns out to be a less desirable specimen. Males of most species, and, doubtless, men of *Homo sapiens*, have always had sex as much as they could, but few could afford multiple wives, just as relatively few could today, even if society permitted it.

Considering the praise that is often heaped on long-term monogamists (usually on such occasions as a golden wedding anniversary), it is noteworthy that monogamy has long been a strategy for the humble, rather than the privileged and successful. Recall the observation, "Those who can, do; those who can't, teach"—an undeserved calumny on a noble profession. But in the world of most animals, those who can have multiple mates, do; those who can't, get by with just one.

If they're lucky.

Most people are surprised to learn that human beings are biologically primed for polygyny, and that harem-keeping was almost certainly the preferred sociosexual "lifestyle" of most of our ancestors during most of our evolutionary past. Men, in particular, are then inclined to express chagrin that they weren't alive to enjoy those seemingly blissful times. What they don't realize is that for every successful harem-keeper in a polygynous species who was mated to, say, N females, there are $N - 1$ *un*mated males. If there are equal numbers of men and women (which is nearly always the case), then for every man with, say, ten wives, there must be nine resentful bachelors. For a man to assume that he would be the one in ten to hit the polygynous jackpot is to commit the same fallacy as those charlatans who claim to recall their previous lives and who invariably announce that they were Napoleon or King Tut, whereas the overwhelming likelihood points to the millions of unsung bits of 19th-century cannon fodder, or an anonymous slave laboring at the pyramids, rather than the Egyptian pharaoh who presided over their construction.

None of us lives in Lake Woebegone, where "all the women are strong, all the men are good-looking, and all the children are above average."

Let's turn now to sexual monogamy as distinct from its socially designated counterpart. Here the situation is more obscure, because there are few topics as prone to confabulation as sex. In general, men tend to overstate their number of sexual partners, in accord with social expectation. Approval, or at least admiration, of the "ladies' man" or "man about town" may in turn have its own biological underpinnings. It has been shown that among certain fish and bird species, females who initially disdain a given male show newfound sexual interest if his situation is experimentally modified so that he is pursued by other females. Attractiveness to others is itself attractive: One can imagine prospective mates wondering whether others might be seeing something that they have missed, and figuring they might as well get on board. Perhaps, in short, there has long been a payoff to men who merely claim to be successful with women; if so, then those who make such professions would have been rewarded, not just evolutionarily (if doing so brought them more sexual success), but also by social acclamation, once cultural traditions began to track biological payoffs.

It appears that women, for their part, typically understate the number of their sexual partners, once again in accord with society's expectation. They are supposed to be chased, yet remain chaste (or at least not to flaunt their sexuality). Even in 21st-century America, which claims to be sexually liberated and egalitarian, men who "play the field" are generally tolerated or even admired (so long as they aren't already married), while women who behave similarly risk being criticized as being "easy" or "sluts." This particular version of the double standard may also be biologically based. For one thing, a woman identified as having had multiple sexual relationships may well be perceived as a potential threat to other, already mated women (the proverbial "home-wrecker"). In this regard, in fact, she is indeed a threat to already mated women, just as the sexually available and attractive man is to other, already mated men.

Biology has another reason to whisper to a woman's unconscious, urging her to downplay her sexual scorecard. It is a consideration unique to women, or rather, to the females of any species that experiences internal fertilization. As we have seen, when fertilization takes place within the female, males cannot be certain that they are the genetic father of any babies that subsequently emerge. Accordingly, females who revealed themselves to be sexually less than faithful were likely seen as maritally less than desirable, even as they may have long been sought after for brief encounters.

We therefore very much doubt that the double standard of self-report—with men claiming more sexual partners than do women—is purely a result of arbitrary social tradition. Here, as with so many other human behaviors, cultural expectations seem to piggyback on underlying biology. In any event, note that the basic evolutionary payoffs, which pretty much qualify as what anthropologists call "cross-cultural universals," need not be consciously understood in order to function, just as one needn't understand the complex nutritional metabolism of food in order to eat when hungry. Once upon a time, people couldn't track their dietary input of calories, vitamins, or good versus bad cholesterol, just as before DNA fingerprinting came along, each sex could practice deception without detection.

And the overwhelming likelihood is that they did so: Men sought, and, when possible, obtained, clandestine EPCs with women who may or may not have been married (which would include women who were polygynously mated), while women accommodated them—and, more to the point, themselves. Thus, while men were often practicing overt polygyny and occasional adultery, women were almost certainly engaging in covert polyandry, enjoying multiple mateships on the sly. As George Bernard Shaw put it in his play *Man and Superman*, "The maternal instinct leads a woman to prefer a tenth share in a first-rate man to the exclusive possession of a third-rate one." The evidence for this is less in-your-face than that for male-structured polygyny, but persuasive nonetheless.

The accumulated data are quite clear, for example, that women partake of a complex pattern of sexuality, in which they not only experience enhanced libido when ovulating (and therefore fertile), but that during this time, they are especially responsive to male traits that are "androgenized," and which therefore signal that their possessors are likely to produce offspring who will, in their turn, be especially attractive to the next generation of women. During the rest of their cycle, when women are much less likely to conceive, they are more likely to be turned on by other male traits, such as wealth, social status, and attentiveness, which suggest a good provider . . . just not necessarily of sperm.

According to P. J. O'Rourke, "There are a number of mechanical devices which increase sexual arousal, particularly in women. Chief among these is the Mercedes-Benz convertible."

One way to make sense of female sexual and romantic preferences is that women have been selected to choose good, reliable providers and cooperators as their social mates, while also being on the lookout for sexy studs to be the genetic fathers of their children. Not that this pattern would necessarily have operated all the time, but probably often enough to leave a distinct evolutionary footprint.

In this regard, testicles have a tale to tell.

Gorillas, for all their large body size, have comparatively tiny testicles. The testicles of chimpanzees, by contrast, are immense. The reason for this difference seems clear: Gorilla males compete with their bodies, not their sperm. Once a dominant silverback male has achieved control over a harem of females, he is pretty much guaranteed to be the only male who copulates with them. Chimps, by contrast, experience a sexual free-for-all, with many different males often copulating in succession with the same adult female. As a result, chimpanzee males have been selected for producing large quantities of sperm, whereby each chimpanzee attempts to maximize the chance that his pollywogs—not those of a competitor—will do the fertilizing. Among most species, the ratio of testicle to body size is a good predictor of the intensity of nonmonogamy: when more than one male is likely to copulate with the same female, the resulting sperm competition selects for the production of lots of these tiny spermatic gladiators.

How, then, do human beings rate in this regard? The testicles of *Homo sapiens* are, relatively speaking, larger than those of gorillas but smaller than those of the champion chimps. The most likely interpretation? Human beings are less certain of sexual monopoly than are gorillas, but not as promiscuous as chimps. Another way of putting it: We are (somewhat) biologically primed to form mateships, but at the same time, adultery is not altogether a dark stranger in our evolutionary past.

Further evidence for "cryptic polyandry" is found in human reproductive anatomy. The uterus, for example, contains numerous cervical crypts in which sperm can be stored for several days, thereby permitting sperm obtained during mating with different men to compete—almost literally to duke it out, mano a mano—therefore enabling the best sperm to win. This situation resembles that observed in our close relatives, the chimpanzees and bonobos, among whom polyandry is dramatically *un*-cryptic. Bonobos in particular have captured the public imagination as the ape that makes love, not war, but in fact they partake of a kind of hidden spermatic warfare, abetted by the polyandrous inclination of their females. To an extent not widely recognized, women, too, may well harbor an inner bonobo.

The "inner" aspect of such inclinations is not surprising; indeed, it is to be expected, given that males in general and men in particular are prone to look with disfavor not just upon sexual competitors (other men) but also upon women who make it obvious that successful competitors will be rewarded with sexual favors. Hence, natural selection may have taken these inclinations on the part of women and internalized them in the literal sense of keeping certain traits under wraps.

Take concealed ovulation.

Women, as it happens, pose a number of evolutionary sexual mysteries—interestingly, far more than do men.[9] Among these enigmas is the fact that ovulation is kept hidden. Unlike chimpanzees, for example, or bonobos, among whom peak fertility is flagrantly advertised by a conspicuous genital swelling, a woman's exact egg release is a closely guarded secret. Even in our high-tech, medically sophisticated 21st century, it is devilishly difficult to tell when a woman is ovulating. Such information would seem biologically important (indeed, crucial), and, moreover, our close relatives clearly are not shy about announcing their own potential reproductive status. This makes it most unlikely that ovulation among *Homo sapiens* is simply inconspicuous, like, for example, the secretory condition of the human pancreas. Rather, a woman's ovulation is actively obscured.

But why?

The short answer is that no one knows, although many hypotheses have been proposed. Among them are several that relate to cryptic polyandry (which, lest it be forgotten, is really a fancy bio-literate way of saying "hidden female penchant for multiple sexual partners," the flip side of what is not so cryptic among males). Note that males of other species are generally selected to concentrate their sexual attention on females who are fertile. This includes both seeking copulation and fending off other males who are seeking copulations of their own. When females oblige by developing a bright red, excruciatingly obvious sexual swelling at these times (think, once again, of chimpanzees or those even sexier bonobos), males respond by copulating with them, and then paying them little attention during the rest of their cycle.

It doesn't do female chimps or bonobos any harm if males know that they've been copulating with other males. Indeed, they may actually benefit, insofar as this induces additional males to, in effect, purchase their own ticket in the reproductive lottery by copulating with females who advertise their availability. Females therefore get a variety of sperm, which then compete among themselves. Just like a horse race with many in the running instead of just a few, a larger number of entries makes it more likely that an especially competent contestant will come out on top. In addition, according to the "infanticide insurance" hypothesis, females get to increase the pool of potentially tolerant father figures.

There are a number of ideas purporting to explain why human beings, unlike chimps and bonobos, are so secretive about their ovulation. One is that since baby *Homo sapiens* are so helpless and require so much parental assistance for so long, women derive a huge payoff from having a contributing male partner when it comes to childcare. Recall that in those species in which males make a genuine contribution to rearing the offspring—notably, birds—males who have reason to think they have been cuckolded are prone to abandoning those offspring. So it may well have been maladaptive for ancestral women to make their sexual availability to other males as obvious as do chimps and bonobos. By keeping their ovulation cryptic, our great-great-grandmothers could have been increasing the likelihood that our great-great-grandfathers would be devoted dads, committed to what they perceived as their own offspring, while at the same time, our cryptic female ancestors would have been capable of mating, on the sly, with other males.

These "other males" would then not only have been recruited into the ranks of the likely noninfanticidal, but would also, of course, be contributing their sperm into the potential donor pool, from which the most desirable would be selected inside the female's reproductive tract. In short, the peculiar human trait of cryptic ovulation may well be mute testimony to cryptic polyandry.

Nor is evidence provided only by women's bodies. The human penis also testifies to the reproductive inclinations of women. If you were to ask a child, "What's a penis for?" he would likely reply, "Peeing." Ask an adult—who would presumably know that women urinate quite nicely despite their penis-less-ness—and he or she would doubtless refer to that organ's role in sex, or, if more biologically inclined, in copulation, specifically, transferring sperm to the female reproductive tract. But in fact, there is another function served by many (perhaps most) penises throughout the animal world: dislodging sperm introduced by other males.

Before copulating, the male damselfly (not quite the oxymoron that it sounds; there are male damselflies as well as females) uses his penis as a scraper, deftly removing sperm deposited by whoever came before. And the damselfly is not unique. Insects generally possess a remarkable penile tool-kit, veritable Swiss Army knives of reamers, scrapers, scissors, levers, ratchets, hatchets, corkscrews, and rasps, all intended to give their possessors an advantage over other males. Some sharks, not to be outdone, have a double-barreled penis, one tube functioning to provide lady sharks with a high-pressure saltwater douche, thereby sluicing away a competitor's sperm before introducing their own.

Among mammals, most species produce as part of their ejaculate something known as a "seminal plug," which hardens into a rubbery barrier that not only keeps sperm from leaking out, but also makes it more difficult for sperm from other males to get in. (Females, not to be outdone, are often adroit at removing them.) Human beings don't produce anything comparable, but there is considerable evidence that men unconsciously adjust their semen as a function of whether they are likely to be competing with other men. Thus, sperm production per ejaculation is higher during sexual intercourse than during masturbation, and higher yet when their partners are more likely to have had sexual relations with other men.

Speculation also abounds that the human penis itself has evolved, at least in part, as an organ of sperm competition. Consider the penile tip, known as the glans. Its peculiar arrow-shaped anatomy is not readily explained, except as a device employed—again, we hasten to add, without any conscious intent—to dislodge sperm deposited by a partner's previous lover. Even the vigorous thrusting so characteristic of human sexual intercourse has been interpreted as facilitating a hypothesized "plumber's helper" role, in addition to simply stimulating ejaculation.

Let's state the obvious: All this huffing and puffing on the part of men, not to mention their presumed structural, penile adaptations, wouldn't have been selected for if women weren't inclined—at least sometimes—to depart from monogamy.

When it comes to cryptic polyandry, all this smoke emitted by both sexes makes it highly likely that there's quite a lot of fire, too, or at least a steady smolder. When not a burning issue, it is guaranteed to be a hot potato. Change the metaphor to a court of law: Hear ye, hear ye, in the case of monogamy versus multiple partners for women, the preponderance of evidence points to multiple partners. Ditto in the case of men, for whom the jury can if anything be even more confident of its verdict, approaching "beyond a reasonable doubt." Add to this the incriminating fact that sexual jealousy is highly developed in both sexes, and ask why, if sexual cheating weren't a high risk and frequent occurrence in our evolutionary past, natural selection would have bequeathed to our species such profound intolerance of a partner's infidelity. By contrast, people aren't predisposed to react with rage (chagrin, maybe) if their lover suddenly becomes invisible or sprouts wings—because this has never happened in our evolutionary experience and therefore isn't "naturally" defended against.

The court of bio-logic, having heard the case, must conclude that monogamy is under siege.

Considering how much has been learned about non-monogamy and EPCs among animals, and considering the current availability of DNA testimony, not to mention the additional evidence just presented, it is remarkable how rarely genetic paternity tests have been run on human beings. On the other hand, in view of the inflammatory potential of the results, as well as, perhaps, a hesitancy to open such a Pandora's box, maybe the widespread human reluctance to test for paternity is sapient indeed. Even prior to DNA fingerprinting, blood group studies in England found that the purported father is the genetic progenitor about 94 percent of the time; this means that for roughly six out of a hundred people, someone else is. And of course, these numbers cannot help but understate the actual frequency of extramarital sex, most of which presumably does not result in pregnancy.

Between 24 and 50 percent of American men report having had at least one episode of extramarital sex during their lifetimes. The numbers for women are a bit lower, but in the same ballpark. On average, between 1.5 and 3.6 percent of married persons report having had an additional sex partner during the past year.[10] Many people already know quite a lot—probably more than they would choose—about the disruptive effects of extramarital sex. It wouldn't be surprising if a majority would rather not be informed about its possible genetic consequences, extramarital *fatherhood*. Maybe ignorance *is* bliss.

But in an age of evolutionary science, combined with a social trend toward increased sexual openness and transparency, that bliss isn't likely to last.

One consequence of this new combination has been to err at either extreme: blaming it all on biology ("My genes made me do it") or to flat-out deny or seek to defy human nature altogether ("Thou shalt not commit adultery"; case closed), with no further questions asked, no nuance, no insight admitted. Just blaming "human nature" is too simple, as well as a recipe for despair (or permission to transgress). At the same time, just saying No just doesn't work.

Indeed, nearly everyone has experienced the pulls and pushes of extramonogamous temptation: being pulled toward another and/or pushed away from an unfaithful partner. And rare indeed is the person who hasn't been affected, nearly always for the worse, by the extramarital transgressions of someone close to them. Few experiences are more disruptive—emotionally, financially, personally, and also laden with impact that reverberates throughout a family, often for years, if not decades.

The poet Ezra Pound once observed (somewhat self-servingly) that artists are the "antennae of the race." These antennae have long been twitching about monogamy: Mostly, it's failures. If literature is any reflection of human concerns, infidelity has been one of humankind's most compelling themes, long before biologists had anything to say about it. The first great work of Western storytelling, Homer's *Iliad*, recounts the consequences of Helen's adultery. Then, in the *Odyssey*, we learn of Ulysses' return from the Trojan War, whereupon he slays a virtual army of suitors, each of whom was trying to seduce his besieged but reputedly faithful wife, Penelope. By contrast, incidentally, Ulysses himself had dallied with Circe the sorceress, but was not considered an adulterer as a result. The double standard is ancient and by definition unfair; yet it, too, as we have seen, appears rooted in biology.

Every great literary tradition, at least in the Western world, finds it particularly compelling to explore monogamy's failures: Tolstoy's *Anna Karenina*, Flaubert's *Madame Bovary*, Lawrence's *Lady Chatterley's Lover*, Hawthorne's *The Scarlet Letter*, Henry James's *The Golden Bowl*, and James Joyce's *Ulysses*. More recently, John Updike's marriage novels—not to mention scores of soap operas and movies—describe a succession of affairs. Infidelity has not only been a cornerstone of popular storytelling, but of real-life political scandal, from Ireland's Charles Parnell to the fall of Rev. Henry Ward Beecher in 19th-century America, to Bill Clinton, Jesse Jackson, Newt Gingrich, et al. in the late 20th (not forgetting FDR, Ike, and JFK in the mid-20th—and that's just a sample of the presidents), to Eliot Spitzer, Kwame Kilpatrick, John Edwards, etc., in the early 21st—with no end in sight.* Nor are women immune, from Russia's Catherine the Great to Great Britain's Princess Diana. Monogamy, by contrast, doesn't seem to generate nearly the attention that is regularly showered upon its failure. It just doesn't seem likely, somehow, that the following headline will make the nightly news: NEITHER OF THE HIGHLY ESTEEMED PUBLIC FIGURES,

* Note, incidentally, that nearly all the renowned cases of adultery involve men. Is this because men are more infidelity-prone, because women are better at hiding their EPCs, or because throughout most of history, men have simply been more famous than women, so their sexual peccadilloes—once revealed—are more attention-getting?

MR. AND MRS. WELL-KNOWN AMERICA, WERE REVEALED TO HAVE COMMITTED ADULTERY TODAY. But maybe it should.

Where are the great novels, songs, movies, plays, operas, poems, blogs, twitters, and other public sagas that explore the painstaking daily effort of one-to-one interpersonal accommodation and cooperation? Happy-ever-afters supposedly occurred in Camelot. (Or not. Ask Guinevere and Arthur about Lancelot.)

Note: The biology of infidelity does not justify it, for either sex, just as the widespread and "natural" inclination to punch an annoying loudmouth in the nose does not condone assault. It may well be that because of human biology, no healthy adult can avoid the temptation to, as Jimmy Carter once confessed to *Playboy* magazine, "commit lust in my heart." The trick, for would-be monogamists, is to keep it there. Insofar as the challenges to monogamy are deep-seated in human nature, however, it becomes especially important for each person to understand the issues, if only to better prepare him- or herself to defy biology's predilections (should this be one's predilection), as well as to better comprehend those of one's partner.

As G. K. Chesterton once observed about Christianity, the idea of monogamy hasn't so much been tried and found wanting; rather it has been found difficult and often left untried. Or at least, not tried for very long. Thanks (no thanks!) to biology, nearly everyone who tries monogamy finds it trying indeed.

Put all this together and the conclusion seems inevitable: Monogamy is biologically contraindicated for human beings, impossible to achieve, and, in any event, not really a good idea anyhow. Right?

Wrong.

To be sure, monogamy is indeed a challenge. Yet it is nonetheless achievable, because even though human biology makes monogamy "unnatural," it also holds keys to its attainment. As to monogamy being impossible, consider that there are myriad monogamous couples occupying this planet, at this very moment, even *happily* monogamous couples. And presumably there always have been. We believe it was polymath Kenneth Boulding who first propounded the following law: Anything that happens is possible.

We hope, in addition, that we have already convinced you that just as being natural does not assure being good, "unnatural" does not necesarily mean bad—sometimes, in fact, exactly the opposite is the case. We are inclined to go further, and suggest that people are never so human as when they act counter to some of their biologically given inclinations. This may be what truly distinguishes human beings from other animals. Even if this proposition proves invalid, we are about to demonstrate—or at least attempt to demonstrate—that there is much in human biology that accords with monogamy after all.

Before examining this positive side of the biology/monogamy connection, we—Barash and Lipton—would like to clarify our personal position and preferences. We are in fact married, to each other, and have chosen to be monogamous. Not in response to the sanctimonious and often hypocritical preaching from right-wing-nut religious ayatollahs, nor as genuflection to current Western social tradition and expectations, but simply because we have decided that, all things considered, monogamy works best for us. At the same time, we are not crusaders for monogamy, or, indeed, for any other particular type of domestic relationship. Our embrace of monogamy—and of each other—is ours, not necessarily yours, and we assuredly are not prescribing it or any other particular lifestyle for anyone else. But if you are interested in delving into some interesting biology—either as a possible guide to "natural" human monogamy, or simply for the joy of finding out how other species organize their love lives—read on.

PARENTING

GIVEN THAT SUCCESSFUL MONOGAMY requires people to say No to some of the most deep-seated human inclinations, part of the miracle of monogamy is that it occurs at all. Indeed, one is hard-pressed to avoid concluding that when it comes to monogamy, biology is baggage. Heavy baggage. But the California mouse can do it, and ditto for the Malagasy giant jumping rat (who you'll meet in the next chapter).

Let's begin, therefore, with a big question: What's so good about monogamy, anyway? For all its drawbacks, and all the pressures that induce living things to depart from monogamy, it keeps popping up—not often, mind you, but persistently and often unexpectedly nevertheless. On a purely scientific and strictly biological level, why does it exist at all? And what, if anything, can would-be human monogamists learn from it?

As it happens, there's quite a lot that is good about monogamy. Not just ethically good, but good when it comes to down-to-earth, evolutionarily adaptive, bona fide biological benefits, genetic payoffs that induce certain nonhuman creatures—who don't take their marital marching orders from gurus, senseis, ayatollahs, ministers, rabbis, priests, or other self-proclaimed purveyors of moral rectitude—to stick with each other. In these cases, as with all animal examples, monogamy exists if and only if practitioners leave more copies of their genes (including any monogamy-promoting ones) than would those who stray into polygyny or polyandry. Animal monogamy can thus be seen as more biologically "genuine" than its human counterpart, since people who practice monogamy may well be responding to the urgings of others, whereas animals simply do what their biology dictates.*

Let's start with biparental care. Although strict sexual monogamy is indeed rare in the animal world, the reality is that certain species—notably, birds—are more likely to be monogamous than are members of any other animal group. This is largely because young birds grow rapidly, and are often helpless at hatching; therefore they require a lot of parental care. It is very difficult to be a single parent when your offspring are vulnerable and needy. (Consider, for example, that many nestling songbirds must be provisioned with an insect every 30 seconds or so!) Not surprisingly, therefore, there is often at least the pretense of willing two-party collaboration. In addition, since genuine monogamy carries with it the promise of genetic parenthood for both partners, fidelity works partly as a lure to get two adults to commit to rearing their shared offspring.

In her delightful book, *Dr. Tatiana's Sex Advice to All Creation*, evolutionary biologist Olivia Judson wrote that true monogamy with sexual fidelity is "so rare that it is one of the most deviant behaviors in biology." But it happens. And when it does, the care of offspring tends to loom large. Let's turn to the mammals and consider, therefore, a revealing case of mice and men, in which monogamy and shared parenting are closely entwined. Such examples, entertaining in their own right, may or may not be inspirational (*March of the Penguins* redux), depending on personal preference. In any event, they can also shed biological light that helps illuminate the human situation.

Mice are not normally considered paragons of monogamy, but there are a few interesting exceptions. Moreover, California doesn't

*As we already suggested, however, there is an important caveat here: Since human beings may well be unique in the natural world in being able to act *counter* to their biology, one can argue that they are most uniquely human when they do so.

typically spring to mind when looking for examples of such supposedly traditional values as bourgeois monogamy, but here is a double exception: the California mouse (scientific name *Peromyscus californicus*), a tiny rodent that is not only socially monogamous but also sexually faithful. Monogamy is generally very rare among other closely related species in the genus *Peromyscus*, known collectively as deer mice. Yet DNA fingerprinting studies have yet to reveal a single case of sexual *in*fidelity in the California species.

These *Peromyscus* just aren't promiscuous. Not only are there no extra-pair copulations, but—equally remarkable—males spend about as much time with their offspring as do females.

Actually, the likelihood is that these two natural history facts are connected: The absence of EPCs is a major reason male California mice spend so much time with their offspring. It isn't just the males who are sexually faithful, but the females, too, and therefore, once a husband California mouse has evolutionary confidence that his offspring are in fact his,* the stage is set for paternal behavior that rivals its maternal counterpart.

Father (left) and mother (right) California mice, with their offspring under them.

PHOTO BY DAVID J. GUBERNICK, WWW.RAINBOWSPIRIT.COM

* Don't be put off by the spurious complaint that these animals couldn't possibly understand the evolutionary genetics of parenthood. It doesn't matter what cognitive assessment—if any—has taken place; all that is needed is that natural selection has conducted the evaluation, rewarding individuals (and their genes) who responded in ways that enhanced their fitness, and selecting against those who behaved in a way that proved genetically destructive.

In a laboratory study, the sexual fidelity of mated pairs was tested by exposing each partner to a new and sexually receptive individual of the opposite sex, while the "husband" or "wife" was nearby but tethered, so he or she couldn't interfere. When tempted in this manner, only two of thirteen males mated with a sexually experienced, estrous female; interestingly, a somewhat higher number of females (15 to 20 percent) mated with newly encountered males. These numbers were unaffected by the presence or absence of a tethered partner, and when given the choice, both sexes strongly preferred to mate with their previous partner.[11]

In all probability, sexual fidelity on the part of female California mice has facilitated paternal care on the part of males, since male childcare efforts are more likely to be genetically rewarded than if expended on behalf of a lady mouse who is fast and loose with her sexual inclinations. Let's grant, in addition, that the benefit of obtaining paternal assistance has selected for such female self-restraint, insofar as sexually faithful females establish a circumstance in which males can invest with confidence in their offspring, knowing that they are really theirs.

But why should these males restrict their own sexual horizons, and what, precisely, do they contribute to the care of their offspring? After all, a notable characteristic of mammals is that females—and females alone—are specifically adapted to nourish their babies, unlike birds, for example, among which males are no less capable than females of provisioning the hatchlings.

Among California mice, having a male around results in substantially greater survival of the offspring. Why? Probably because the male is a genuine cuddler, and his physical presence helps keep the pups warm, and thus alive. For about two weeks after their birth, infant California mice are essentially cold-blooded. Not only that, but they appear in unusually small litters for a mouse—just two at a time—and they do so at the coldest season of the year. The newborns cannot maintain their body temperature, and there aren't enough babies to keep each other warm. This is where Daddy is especially inclined to lend a paw—or rather, a warm body.[12]

It's also noteworthy that even when males of this species have territories large enough to encompass the living space of more than one female, they don't do so. It may well be that male California mice are reluctant to spread themselves—and their pup-preserving body heat—too thinly.

There appear to be two different kinds of male California mice, and they differ in their fatherly inclinations: Most of them start acting paternal only after the young arrive, and are stimulated to do so by chemicals produced in their mates' urine. A not-insignificant minority, however (about 25 to 30 percent), are prepared to act like doting fathers while their "wives" are still pregnant; this has been tested by exposing a male mouse to unrelated newborn babies (at any stage of his mate's breeding cycle). Paternally inclined males begin behaving paternally right away; the others do so, too, but only after their own offspring are born.[13]

Human beings, like California mice, are unusual in the degree to which men provide paternal care. In California mice, fathering is very much a manly mouse thing: When males are castrated, their paternal care plummets, although, interestingly, their tendency to be aggressive toward other males is not reduced. So testosterone is a guy thing among California mice, but it leads to being a doting father, rather than a doughty fighter. In the world of *Peromyscus californicus*, at least, it's not that mousy men take care of their kids, but rather, manly mice are the ones that do so.

Like California mice, human beings are unusual—certainly exceeding all other primates—in the degree to which men act parentally. Bear in mind that mammals are generally at the very low end of the spectrum when it comes to biparental care—probably due to the mammalian combination of internal fertilization (hence, low male confidence of genetic relatedness to offspring) and female lactation—hence, little in the way of built-in male capacity to contribute, compared to a nursing mother.*

Not coincidentally, human beings are also helpless at birth, and they remain dependent for years, sometimes decades. Biologically, like all mammals, our species is primed to receive nutrition from our mothers, and although human babies don't necessarily need a father's physical warmth, human fathers benefit their offspring in many ways: physically, emotionally, financially . . . all adding up to greater success for the children and, as a result, for the parents as well. Clearly, single-parenting is possible (Boulding's Law, once again), but it is basic biological common sense that two are better than one. In numerous studies of nontechnological societies, anthropologists have found that children from "intact" families have higher survival rates and are, therefore, more likely to contribute to the evolutionary success of their parents.

In the United States, sociologists agree that children from two-parent families are more likely to graduate from high school and college, less liable to have run-ins with the law, more likely to be physically and emotionally healthy, and so forth. Staying together "for the good of the children" may not make the heart sing, and isn't necessarily a good idea in all cases. Biologists have moved far beyond "the good of the species" arguments, since individual and gene-focused payoffs are what drive evolution, so an argument that people should stay together for the good of the species makes even less sense. But there seems little doubt that, at least on occasion, "the good of the children" can be an enormous individual and gene-focused benefit of staying together—and thus, of monogamy.

* A question not often asked is, "Why don't men—or other mammals—lactate?" After all, once the female has undergone the stresses of pregnancy and childbirth, wouldn't it be fair for the male to pitch in and produce milk? The answer, almost certainly, is that fairness doesn't count in evolution; genetic success does. And males couldn't be expected to lactate, like females do, unless they were absolutely guaranteed relatedness to their offspring . . . like females are.

A quick look at parental payoffs suggests a useful way of considering the whole monogamy enterprise: game theory. Game theory is often dauntingly complex and highly mathematical, but it needn't be.[14] At minimum—and "minimum" is precisely our goal in the coming pages—it simply helps conceptualize interactions between two "players," say, a man and a woman. When their actions are independent, there is no "game." For example, if it is raining, either or both or neither may choose to carry an umbrella. Depending on the choice, the decider may or may not get wet (and may or may not be unnecessarily burdened if the rain stops), but neither payoff depends on what the other does. Nor does either decision influence the weather itself. In such cases, evolutionary biologists expect individuals (of any species) to behave in a manner that maximizes their benefits, minimizes their costs, and doesn't require a whole lot of sophisticated analysis.

Things get interesting, however, when the potential payoff to each player depends not only on what he or she does, but what the other one does at the same time. That's when game theory applies. Imagine, for example, a male and female (of any species) that are about to mate, or have just done so. Each must decide, whether consciously or un-, if it should stay with the other and assist in rearing any offspring that may result. Here is a simple "payoff matrix," in which both the male (displayed horizontally) and the female (vertically) can either Stay or Leave. The resulting four outcomes are shown inside each square. When both Stay, the result is monogamy (upper-left corner); if the male Leaves and the female Stays (upper-right corner), the result—commonly found in such birds as mallard ducks, or mammals such as bears—is that the female ends up a single parent. When the male Stays and the female Leaves (lower left), this is the situation found in some fish and frogs, and when both parents Leave (lower right), no one does any childcare—this is the situation of oysters, for example, in which males and females simply squirt eggs and sperm into the water, leaving the resulting offspring as pelagic orphans. Or, if the babies require parental care, it is provided by an unwitting victim—typically

As for traditional internally fertilizing species, such as birds and mammals, the situation is quite different. A male can inseminate a female and then leave, whereas she cannot very well run away from her own developing fetus. And not surprisingly, the world of birds and mammals is rife with cases of deadbeat dads and single-parent moms. But there are monogamists nonetheless, species that appear in the payoff matrix's upper-left corner. Summarizing a complex situation, it only pays males to leave (upper-right corner) if they can enhance their evolutionary fitness by doing so, which means that they must have some prospect of achieving additional copulations, *and* their offspring must be able to do comparatively well with only their mother's assistance. When infants are helpless at birth or hatching, and if they require so much parental aid that a single mother is unlikely to do as well as two devoted parents, the cost-benefit calculus shifts and converges on the upper-left corner—monogamy.

Human beings score high on both these considerations. Hence, although monogamy isn't mandatory in our species, it has much to recommend itself. Lets look, therefore, at a very simple parental payoff matrix for human beings:

		MALE'S BEHAVIOR	
		Stay	**Leave**
WOMAN'S BEHAVIOR	**Stay**	Monogamy	Deadbeat Dad
	Leave	Deadbeat Mom	Abandoned Orphans

The human situation, of course, is far more complicated than this. A woman—or man—may end up as sole childcare provider because his or her former partner is abusive, dangerous, or otherwise incompatible or inappropriate. Either parent may leave for greener pastures, or for any number of reasons. But there can be no doubt that among *Homo sapiens*, no less than *Peromyscus californicus*, the neediness of children can lead to steadiness of shared parental devotion. Or, putting it another way, ancestral parents who were somewhat monogamously inclined are likely to have (upper left) more descendants, who in turn will have a penchant to keep doing so, if given half a chance. Even devoted parents who were successful at EPCs—which, by definition, took place "on the side"—would likely provide more consistent child support than those who defected, and more consistent child support leads to more successful offspring.

Let's take one more look at parentally motivated monogamy, employing a tiny bit of game theory. When theoreticians consider games like those shown in the payoff matrices presented so far, they can be considered variants of the famous (or infamous) Game of Chicken. Consider a classic Game of Chicken, in which two people drive their cars toward each other, straddling the white line, each daring the other to swerve aside. Whoever does, loses. The one who goes straight while the other swerves, wins. If both swerve, then both come out ahead, although each gains somewhat less than if only the other one had swerved. If both go straight, then both get the worst outcome of all, a head-on collision. A payoff matrix for a Game of Chicken would therefore look like the diagram on the following page.

Games of Chicken are notoriously dangerous, especially since the lure of a victor's payoff—getting the other driver to swerve, and thus emerging as some sort of courageous, determined hero*—is so great as to motivate each driver to take the chance that the other one will swerve. Substitute "Leave" for "Go Straight" and "Stay" for "Swerve,"

*Whereas, in fact, most objective observers would consider the winner (and, indeed, anyone who plays the game) to be a demented lunatic.

		DRIVER #1	
		Swerve	**Go Straight**
DRIVER #2	**Swerve**	Both Win	#1 Wins
	Go Straight	#2 Wins	Both Lose

and reduce the personal costs of a mutual collision, and we have a version of the "monogamy game." Thus, the lowest payoff for both parents is liable to arise if each one abandons his or her parental duties, and the offspring become orphans. It might nonetheless be tempting to attempt such desertion, however, if there is sufficient likelihood that the other parent will rise to the occasion and take up the childcare slack, leaving the "winner" free to pursue additional possible prospects.

But there is a much more appealing, altogether feasible solution to this reproductive "game." If the payoff of shared parenting (back to that upper-left corner once again) is high enough, and the cost of abandonment is severe enough—not only in the event of both parents leaving, but even just one—then both players would be well advised to swerve . . . right into each other's arms. If so, everyone wins. Including—not least—the children.

It hardly matters, incidentally, if would-be monogamists are childless (or, as many prefer to be designated, "child-free"), or if they currently occupy an "empty nest," or if they are homosexual rather than heterosexual. The overwhelming thrust of evolution is provided by what our ancestors have been doing—and what has worked best—during our overwhelmingly ancient past. And one thing our great-great-grandmothers and -grandfathers were doing was rearing children: Everyone alive today is the product of an unbroken line of breeding adults who reproduced without ever missing a beat, way back to the prehistorically parental primeval porridge. So reproducing is in our blood, our genes, and our brains, even in modern times when people finally have the option of choosing not to. Thanks to birth control, we can decide to refrain from reproducing,* but this doesn't mean that our species-wide inclinations still aren't tuned to a "breed-as-much-as-you-can" world.

Nearly everyone finds a potential sexual partner attractive in direct proportion to how much he or she is relatively youthful, physically healthy, and temperamentally pleasant. (If the prospect of reproduction didn't underpin romantic preferences, there's no reason why people wouldn't be as turned on by a 90-year-old crone as by a 20-year-old *Playboy* centerfold.) And similarly, when it comes to the prospect of a lasting mateship, our biology has outfitted us with a soft spot for others who show themselves to have monogamous potential, even if both parties are committed to their own domestic version of zero population growth.

*And, in an increasingly overpopulated world, let's hope that more and more people make that choice!

Chapter 5

RECIPROCITY

THE BENEFITS OF SHARED PARENTING, although important, don't exhaust monogamy's possible biological payoffs. An important one is reciprocity, i.e., one good turn deserves—and receives—another. You scratch my back, I'll scratch yours. You be around for me, and I'll be around for you. And so, let's meet the Malagasy giant jumping rat.

The likelihood, however, is that you never will, since this species is only found on the island of Madagascar and is endangered—which is a shame, at least in part because the Malagasy giant jumping rat also shows a behavior trait that is rare: monogamy. (So, too, does another endangered creature and from Madagascar: the fat-tailed dwarf lemur, about which even less is known.)

Fat-tailed dwarf lemur.

PHOTO BY DAVID HARING, DUKE LEMUR CENTER

Malagasy giant jumping rat. PHOTO BY PIOTR LUKASIK

At first glance, the Malagasy giant jumping rat (or "vositse," as it is known to the local Malagasy people) looks like a cross between a small rabbit and an even smaller kangaroo. It is an endearing creature, rabbit-sized, sporting upright, pointy ears and hopping on two sturdy rear legs, using its long, muscular tail for balance while propelling itself up to three feet in the air. As far as can be determined, it is a lifelong monogamist, with male and female remaining together until one of them dies: no divorce, no desertion, and, it appears, virtually no adultery. A typical Malagasy giant jumping rat family isn't especially gigantic, consisting of Mom, Dad, and—unusual for rodents—just one baby at a time.

Like California mice—who also, interestingly, have exceptionally small litters—Malagasy jumping rats aren't only socially monogamous but sexually, too: In one DNA study, only two "extra-pair" offspring were found out of 48 parent-offspring trios that were examined (suggesting that these creatures experience, if anything, more sexual fidelity than do human beings).[15] They are nocturnal, spending their days in underground burrows, then jumping out and about at night. Males and females cooperate in defending their territories from other pairs and from wandering individuals of either sex.

Why are they monogamous? As with the California mouse, the answer seems to be related to care of the young—in this case, not keeping them warm, but keeping them safe. And in this case, it appears that biparental cooperation is key. Thus, we have noted that Malagasy giant jumping rats are exceptionally unrodent-like in having a very slow reproductive rate: Not only do they give birth to just one offspring at a time, but pregnancy lasts more than three months. Each time they breed—which they don't do all that often—these animals have all their eggs in one basket, an offspring who, moreover, is also highly vulnerable, particularly to predators.

The good news for this odd little creature: The Malagasy giant jumping rat is the largest rodent on Madagascar. The bad news: It is the preferred prey of the largest native carnivore, known as a fossa—a mink-like relative of the mongoose and meerkat—as well as ground-dwelling boa constrictors. In all likelihood, the vositse's best defense is to retreat into its elaborate defensive burrow system, and it takes two adults to defend such a structure from other vositses, who would love to appropriate it for themselves. Beyond this, it appears that the presence of an adult male Malagasy giant jumping rat reduces the frequency with which predators make off with the pair's precious Malagasy giant jumping ratling . . . although exactly how the pair actually cooperate to defend their offspring (and, indeed, whether they actually do so) remains unknown.

For an even clearer case of how shared effort leads to pair bonding, we turn to the world's tiniest monkey, a family known as the Callitrichidae ("*cal-a-trick-idee*"), which includes the marmosets and tamarins. These adorable miniature monkeys are found in the New World tropics, and have the unusual trait of generally producing twins as well as being monogamous.[16] Thus far, EPCs have not been reliably documented among any of the marmosets and tamarins. Although it wouldn't be entirely surprising if someday it is revealed that some marmosets and tamarins occasionally have sex with someone other than their spouse, it would be very surprising indeed if this were at all common. And once again, the "miracle" of marmoset and tamarin monogamy is achieved by virtue of their biology.

Perhaps the most notable Callitrichid is the pygmy marmoset, the world's tiniest monkeys, whose infants weigh a mere one-half ounce at birth. Since even adult pygmy marmoset females only weigh a few ounces, a two-baby litter, despite being a mere one ounce of marmoset meat, represents a substantial energy investment. Moreover, nursing is expensive, too: Females can lose up to 22 percent of their body weight during lactation. And this leads us to the likely reason for monogamy among these animals: males pitching in and carrying their babies.[17]

Ecologists use "carrying capacity" as a technical term, referring to the ability of any habitat to support particular numbers of a given species. For the Callitrichids, "carrying capacity" is the literal ability of adults to carry their kids. If you're a very small monkey, this is quite costly, adding as much as 20 percent to your metabolic burden. And this, evidently, is an added cost that a nursing female simply cannot bear. Enter her husband. Female tamarins and marmosets get preg-

Two young pygmy marmosets being carried by their father.

PHOTO BY JULIE LARSEN MAHER, ©WCS

nant, give birth, and nurse, after which males schlep the babies. It is costly for them, too—they lose weight, get to do less feeding, and can't jump as far while carrying their twins (which is probably important when trying to escape from tree snakes), but, unlike their "wives," at least the males haven't already been depleted by pregnancy and nursing. The resulting division of domestic labor is not only heartwarming, but biologically effective. When biologists speak of marmosets or tamarins "carrying on," they are referring not to sexual escapades, but to a devoted father literally carrying his twins on his belly or back.

It is probably no coincidence that pygmy marmosets aren't only the most monogamous of nonhuman primates, but also the species that takes the prize for having the most "mothering males." Daddy marmosets even undergo hormonal changes that prepare them to care for their infants at the time of birth: elevated prolactin levels and decreased testosterone. Testosterone levels in males diminish even further as they engage in childcare. And testosterone levels are lowest in those with the most paternal experience.

Marmoset mothers start disengaging from their twins a few weeks after giving birth. At this point, the male steps in: grooming, feeding, and, of course, carrying his twins, sometimes with help from their older siblings. There are even reports that father marmosets sometimes act as midwives, grooming and licking their babies immediately after they are born and—in one anecdotal account—actually helping pull a baby out of the birth canal.

Devoted marmoset dads, lowered testosterone and all, are not devoted only to childcare, however. Two weeks after she has given birth, a female pygmy marmoset is likely to be pregnant again, her rapid recycling almost certainly facilitated by the fact that someone else—her mate—is now taking over most of the childcare duties. (One is tempted to conclude that in return for good parenting, the pygmy marmoset male is rewarded with good sex.)

There is another wrinkle to Callitrichid monogamy. Although sexual infidelity seems exceedingly rare, there is sometimes room in the social lives of these tiny monkeys for more than a pair—*the* pair—of breeding adults. Thus, marmoset and tamarin families occasionally consist of additional, helper adults. As one might expect, the main function of these extra hands is to help out in the baby-carrying department. These hands usually belong to sub-adult and adult males, with whom, interestingly, the breeding female does not copulate. (Presumably they get thanked some other way, most likely by the fact that they are often genetically related to the breeding adults, so when offspring are produced, these helpers are also helping their own genes get ahead.)

It is also noteworthy that baby-carrying helpers are more frequently found among species that have to travel relatively long distances in order to meet their food requirements. Among the more sedentary types—such as the pygmy marmosets, in which babies are sometimes "parked" in the crotch of a tree while the parents go out to dinner—helpers are much less frequent, and two adults, committed to each other socially, sexually, and reproductively, are sufficient and the norm.

At first glance, it seems easy for cooperative partnerships to form—in the natural world in general and among human beings in particular. Why, then, should cooperative male-female relationships, such as among Malagasy giant jumping rats or pygmy marmosets, be so rare? The reason is actually simple: cheating. Not just sexual cheating—although it is significant that the word is often used to refer to adultery or infidelity—but the more fundamental temptation to take from another, or, more to the point, accept assistance from him or her, without giving back in return.

Cooperation cannot evolve, or persist, if it is not mutual. This is not only a definitional requirement, since cooperation *means* mutual collaboration, but also a biological one, since in the real world, cooperation often requires that someone go first, and essentially donate his or her time, energy, or resources, expecting that the beneficiary will reciprocate. If this happens, then both parties can be better off, as in those cases of monogamous, cooperating mateships.

But along with the allure of shared benefit to cooperating partners, there lurks the prospect of unilateral, selfish gain by unscrupulous (but evolutionarily successful) cheaters.

One of the most important realizations in modern evolutionary biology involves exactly this issue, first developed by Robert Trivers when he was a graduate student at Harvard University. In Trivers' day, the late 1960s, biological theory had recently been rocked by the insights of British geneticist William D. Hamilton, who showed how natural selection actually worked primarily at the level of genes, and how this, in turn, explained much "altruism," the otherwise perplexing observation that living things sometimes behave in a manner that reduces their own reproductive success while enhancing that of others. Animals occasionally share food, assist each other to ward off predators, even help other individuals to reproduce, thereby benefiting the recipients at some cost to themselves. Under the older paleo-Darwinian paradigm, any such altruistic acts should quickly be selected against, since their genetic underpinnings would necessarily be replaced by selfish alternatives, what economists call "free riders."

Hamilton's research showed that by focusing on the level of genes, rather than bodies, natural selection could select for altruism. The quick-and-dirty explanation is that since relatives share genes with a predictable probability (close relatives are simply bodies with a high probability of sharing those genes—the closer, the higher), when one individual benefits another, the "altruist's" genes may actually be benefiting themselves, encased within the beneficiary's body. Hence, what appears altruistic (and thus, counterevolutionary) at the level of bodies can actually be selfish (and thus, consistent with evolutionary theory) at the genetic level. A neo-Darwinian revolution was underway.[18]

But a problem remained. What about seeming altruism between *non*-relatives? Clearly this couldn't be generated by Hamilton's process, which came to be called "kin selection," since the individuals in question were not always kin and therefore did not share identical genes via common recent ancestry. Trivers' analysis plugged this scientific loophole by showing how altruism could evolve even in the absence of shared genes.[19]

The answer was reciprocity: I help you, after which you're better off, and shortly thereafter you help me, so I'm better off, too. Such an arrangement not only appeals to our sense of fairness and propriety, but also warms the heart. For it to work, however, two conditions must be met. Number one: The initial cost to the altruist must be small, and the potential payoff, once the action is reciprocated, must be greater than that cost. Otherwise, it's a losing proposition. In itself, this isn't a terribly high hurdle. Imagine, for example, that you have a lot of food and someone else is nearly starving. You give her some of your surplus (small cost to you, big benefit to her), and then, at a later time when you are in desperate straits, she shares with you. Big benefits all around. Evolution should favor such a system.

But then there is condition number two: The first beneficiary must reciprocate when the opportunity arises for doing so. What is to keep that initial recipient from taking your beneficence and then failing to reciprocate? Such behavior would be ethically unacceptable, but, given the cold-eyed calculus of natural selection, it would likely be favored by evolution, since such cheaters would be somewhat better off than honest reciprocators.

Trivers recognized the problem; indeed, much of his groundbreaking research paper explained it in semi-mathematical form, invoking game theory, specifically a pattern known as the Prisoner's Dilemma. As Trivers demonstrated, in order for reciprocity to evolve it must somehow solve the problem of cheaters, which is to say, it must overcome the Prisoner's Dilemma. And as we are about to argue, the same thing applies to monogamy.

Here is the Prisoner's Dilemma in a nutshell. Imagine that two individuals, #1 and #2, each have a choice: to cooperate or to defect. Cooperators give food; defectors accept food but don't give any. Cooperators swerve their cars; defectors drive straight ahead, come what may. Cooperators are nice; defectors are nasty.* Cooperators make good monogamists; defectors don't. Lets further imagine that individual #1 doesn't know what #2 will do, and vice versa, but as in other games, the payoff to each depends on the other's action, no less than her own. And finally, assume that these payoffs are ranked as follows: The highest payoff accrues to someone who defects (in our example, takes food from a cooperator but doesn't repay it). In game theoretic notation, she receives payoff T, the Temptation to defect (or cheat) when the other player cooperates. The next-highest return goes to cooperators, so long as the other player is also cooperating—this is R, the Reward of reciprocal cooperation, and though it is not as high as T, it is nonetheless quite respectable. Then comes P, the Punishment of mutual defection, endured by both players when each resorts to cheating. And finally is S, the Sucker's payoff, lowest of all: the fate of a cooperator who interacts with a defector, a nice individual taken advantage of by a nasty cheater. When the relationship among the payoffs is $T > R > P > S$, the result is a Prisoner's Dilemma.

		INDIVIDUAL #1	
		Cooperate	**Defect**
INDIVIDUAL #2	**Cooperate**	Both get R	#1 gets T, #2 gets S
	Defect	#1 gets S, #2 gets T	Both get P

*Serious, mathematically minded game theorists actually use the terms "nice" vs. "nasty," as well as "cooperate" vs. "defect."

What's the dilemma, you might ask? Listen in on the logical thoughts of either player as she tries to decide what to do, and bear in mind that the only consideration guiding that decision is how to get the highest possible payoff:* "I don't know what the other individual will do, but I do know this—if she cooperates, my best move is clear: defect. That way I get T, the highest payoff of all (and she gets S, the lowest). On the other hand, she might defect. In that case, what should I do? That's easy: I should also defect, since that way I get P, which is admittedly a punishingly low return, but at least it's better than S, my payoff if I'm a Sucker and cooperate when she's defecting. And so, regardless of what the other player is going to do, my course is clear: defect."

The other player, of course, can be expected to reason similarly. And so, responding to the altogether reasonable Temptation to cheat while also—and again, very reasonably—fearing to be made a Sucker, both players end up defecting and receiving P, the Punishment of mutual nastiness. The dilemma, therefore, is simply this: Given the devilish nature of an interaction that otherwise seems so simple, both players find themselves with a regrettably low payoff (P, the Punishment of mutual nastiness), when each could have received a higher one (R), had they only figured out a way to get the Reward that comes from cooperating. Following the siren song of maximizing fitness, natural selection cannot help leading would-be cooperators into a morass of selfish, mutual cheating . . . which, paradoxically, produces a result that is, in the long run, less self-serving than self-defeating. Evolution giveth and evolution taketh away.

Time now for the good news. There is light at the end of this particular tunnel, one that has been extensively researched by game theorists, economists, political scientists, sociologists, computer scientists, biologists . . . and way too few marriage counselors. The classic Prisoner's

---

* This is precisely what natural selection does, with "fitness" or "genetic success" constituting the only meaningful payoff.

Dilemma assumes, among other things, that there is only one interaction, after which individuals #1 and #2 go their separate ways, never to meet again. When this is the case, there seems no denying that each is obliged to defect. It is doubtless not coincidental that con men are most likely to be "here today, gone tomorrow" types, like the iconic Harold Hill in *The Music Man*, rather than old friends. Or that people are well advised to trust a salesman with whom they have done business in the past and whom they expect to patronize in the future, rather than someone who is just passing through.

In the real world, individuals frequently have the opportunity to interact many times, and when this is the case, they can seek—and obtain—benefits not available via one-time exchanges.* A key consideration is what political scientist Robert Axelrod called the "shadow of the future." In this case, "shadow" is a good thing, and the longer it is, the more likely is cooperation. When a relationship persists, or even if it just has the potential and expectation of persistence, the more each participant can be confident that the other will behave cooperatively. In part, this is simply because—despite what stockbrokers and financial analysts are constrained to say—the reality is that past performance is the best predictor we have when it comes to future returns. It's no guarantee, but it sure beats a wild guess based on nothing whatsoever. And that's why people are more likely to give the benefit of the doubt to a trusted friend rather than to a complete stranger. It's also why longtime friends are more likely to be trusted in the first place.

The relevance of all this to monogamy should be apparent. To be sure, reciprocity can occur among individuals who are not domestic partners, and it often does. Consider such famous partnerships as Laurel and Hardy, Smith and Wesson, Abbot and Costello, Watson and Crick, Currier and Ives . . . all of them productive, but none *re*productive

*Even in the case of repeated (technically known as "iterated") exchanges, a strictly mathematical analysis demonstrates that logic alone still dictates defection (see note 14); here is a case, however, in which human psychology appears to intervene, generating potentially higher payoffs than hardheaded logic alone can provide.

(with each other, that is). Or—our favorites—the creators of great musicals, such as Lerner and Loewe, George and Ira Gershwin, or Rodgers and Hammerstein. These are especially good examples of reciprocal collaboration, based on cooperative, positive contributions in which the strengths of each complement those of the other: say, between a composer (Richard Rodgers) and a lyricist (Oscar Hammerstein). Interestingly, the greatest of all such partnerships—between William Gilbert and Arthur Sullivan—involved two people who didn't much like each other, but whose partnership benefited both. It isn't only sexual partners that can end up making beautiful music together.

More commonly, of course, nonreproductive partnerships do, in fact, involve people who get along . . . our point is that what gets people to get along is the seemingly dreary fact that by doing so, they create something more than the simple payoff available to each alone. It may also sound coldhearted, but in fact, we suspect that even friendships (and not just business partnerships) pivot on reciprocity—not so much "What have you done for me lately?" as "What have you done for me at some time in the past, and can I count on you in the future?" If you invite someone to dinner once, then twice, and no reciprocal invitation is forthcoming, what does that say about the relationship? Once you perceive someone as a consistent taker but never a giver, are you likely to like that person? Friendship, in short, equals reciprocity, and often—although not always—vice versa.

Yet, given the seeming logic of "You scratch my back, I'll scratch yours," clear-cut cases of reciprocity (outside reproductive pair bonds) are surprisingly rare in the animal world. Mateships are a special case of "friendships" generally, further cemented by shared reproductive interests, or, when reproduction is not in the cards, by what psychoanalysts call "libidinal cathexis," which is simply the added closeness that comes from sexual intimacy. And so, mateships offer the added potential benefit of reciprocity that delves much deeper than that available to "just friends," however close.

To be sure, there are relationships that involve the continued Punishment of mutual defection—these are called bad marriages. There are those in which one party consistently defects (and presumably gets T, a payoff measured in alcoholic excess, infidelity, laziness, under-functioning in other respects, and so forth), while the other absorbs the cost, but stays in the relationship (getting the Sucker's payoff, S). These are situations of codependence.

But for anyone in a committed long-term relationship—that is, monogamy—the highest shared payoff comes from being cooperative. At least as important, moreover, is the converse: If you want a whole lot of cooperation, then it pays to be monogamous. Of course, you don't have to be sexual partners to enter into an ongoing pattern of reciprocal benevolence. It happens all the time, and is called friendship. But monogamy is thc ultimatc friendship, which is why the best marriages are between individuals who are not only lovers and often co-shareholders in each other's fitness, but also friends.

A key—more likely, *the* key—isn't so much the behavior of either individual taken alone, but the interaction between them, what physicists might call the "field" that they jointly establish. According to Albert Einstein, "a courageous scientific imagination is needed to realize fully that not the behavior of bodies, but behavior of something between them, that is, the field, may be essential for ordering and understanding events."[20]

We might also take a cue or two from downy woodpeckers. These animals have to worry about predators, notably sharp-shinned hawks, Cooper's hawks, and small falcons known, (incorrectly) as sparrow hawks. Like many other animal species, downy woodpeckers sometimes give alarm calls when they spot a predator. But it's a dangerous thing to do, since the alarm-caller makes himself more conspicuous to the predator. If the goal were simply self-preservation, the most rewarding thing to do would be to keep quiet. After all, you've spotted the danger, and are therefore less vulnerable to it—why not save yourself and let others suffer the consequences of their ignorance?

Indeed, this is what happens most of the time, among downy woodpeckers and most other animals as well. The only consistent exception is with close relatives, among whom shared genes pave the way for substantial altruism, with the "selfishly" altruistic genes promoting their own success in the bodies of those who benefit from the alarm-calling. A study conducted in the Great Swamp National Wildlife Refuge in New Jersey found that, as expected, solitary woodpeckers always kept quiet when they spotted a predator. Ditto for male woodpeckers in the company of other males, and for females temporarily associated with other females. The exception? Mated pairs, who undertook the risk of warning each other of danger, even at substantial possible cost to themselves.[21]

We don't know if downy woodpeckers consider themselves to have friends, whether of the same or of a different sex. But it seems a good bet that the closest friendships of all—and the most reliable ones—are those between mated pairs, who have good reason to count on each other. As Einstein might have put it, there is "something between them."

It may seem strange, or even somehow unethical, to point out these and other selfish payoffs of monogamy—thus far, reciprocity in general and shared parenting in particular (there are more to come). But recall that the biological pressures for *non*-monogamy come from various other "selfish" considerations: for males, producing additional offspring via other partners, thereby enhancing their own success in the process, and for females, increasing the success of the offspring they produce, regardless of who is the father. So selfishness, at least at the genetic level, is a primary motivator in any event, whether acknowledged or not. How appropriate, therefore, that the most powerful forces counteracting our species' antimonogamous, biological predispositions are also selfish, and similarly biological: greater success in rearing children with one's mate, and the numerous benefits of cooperation and coordination from another individual, not to mention the other benefits we are about to consider.

For young people, "friends with benefits" means a friend with whom one occasionally has sex. For older people, it might mean someone who has a good health insurance plan. Or, maybe, one who drives at night. George Bernard Shaw once quipped that youth was so wonderful, it's a shame that it's wasted on the young. The benefits of cooperation and collaboration—with or without sex—needn't be limited to the young. Certainly, it isn't wasted.

In its strictly defined biological sense, reciprocal altruism occurs in only a very small number of animals, leaving human beings as the social reciprocators par excellence. In other words, the biological benefit of monogamy doesn't only involve parenting, but also includes sharing and cooperating, which enables both participants to feel better—and do better—than either would likely achieve alone. It is worth noting, therefore, that even though the evolutionary stage man well have been set by the payoffs of shard heterosexual parenting, one he curtain has risen other benefirs—notably stuff, love and happiess—become available to any adults who seek them. As a result, monogamy can make perfectly good biological sense for the same-sex couple.

Chapter 6

REAL ESTATE AND OTHER STUFF

THE OLD SAYING IS WRONG. Two cannot live as cheaply as one. On the other hand, two together can indeed live more cheaply than the sum total of each, living separately. Cost aside, two can also achieve more, together, than either can alone. In the realm of relationships, especially, a monogamous whole can be greater than the sum of its parts (even beyond the benefits of reciprocity, in which each contributes something that the other may lack, à la Rodgers and Hammerstein) Since we're taking a hard-headed, biologically based look at monogamy, let's not ignore the direct material payoff of cooperation, coordination, and collaboration.

In this regard, we can learn a lesson or two from beavers, justly renowned as natural engineers who construct complex dams and elaborate lodges. Less widely known: Much of their success is built upon monogamy. A typical beaver family consists of one adult male, one adult female, and their offspring of the current and previous years. Although their monogamy is not necessarily lifelong, it is impressive.[22] As with human beings, serial monogamy is the most frequent pattern, with mateships lasting about as long, on average, as they do among people.

Why are beavers monogamous? The most likely explanation has to do with their unusual lifestyle. These animals eat tree bark and other woody plants, all of which are quite low in nutritional value. Beavers therefore need a lot of real estate to sustain themselves and their families, whose numbers are constrained by how much food can be reaped from their immediate surroundings. In addition, beavers typically have to work hard constructing canals in order to float food to their lodges and building material to their dams, as well as to help them escape from predators. And of course, they spend a lot of time not only eating and transporting things, but also constructing and maintaining their homes and the dams upon which they depend to maintain adequate water levels. Put this all together, and beaver homesteading becomes a full-time job, requiring two hardworking adults who are kept as busy as, well, beavers. Too busy to spend time gallivanting about, looking for additional mating opportunities.

Male and female beavers are about the same size (actually, females are a bit larger), and both male and female will actively repel any intruder of either sex. Since their real estate is valuable—largely because of their own "value-added" contribution, a beaver equivalent of "sweat equity"—these animals do best when they support each other in creating a successful homestead. Like hardworking settlers in the American West, male and female beavers are both diligent and cooperative; there is simply too much work for one individual to handle. Male and female take part, pretty much equally, in scent-marking and defending their property, repairing lodge and dam, gnawing down trees and transporting them, as well as constructing canals. When the young are born, however, division of labor develops: Mother does nearly all the childcare, while father ratchets up his outdoor chores.

Beaver kits remain dependent for several years, during which time their parents must continue to maintain the family farm. Young beavers also have a lot to learn: how to cache food in the fall, the fine points of dam repair and construction, what trees are good to eat and which to avoid, and so forth. And of course, home schooling is labor intensive and likely wouldn't be successful if beaver "marriages" weren't equally so.

Staying together because of the material benefits may well be even less appealing than doing so "for the good of the children." But ignoring these benefits, not to mention the cost of forgoing them, can be downright stupid. The old expression "helpmate" deserves new respect, and not only because of the potential payoffs of having a mate who helps; there is also the cost of losing one. Just as solitary beavers typically do not prosper, part of the unmentioned downside of divorce is the literal financial and socioeconomic penalties often incurred. Not that we're saying "Leave it to beaver"; sometimes the lodge, dam, and pond just aren't worth living with a supposed helpmate who is more like a mate from hell. Nonetheless, a bit of beaver-like enlightened self-interest (or, at least, taking time to consider what these creatures have to offer) can give added momentum to monogamy.

And not just beavers. One of the more exotic behaviors of certain bird species—notably concentrated, for some unknown reason, among tropical species—is the phenomenon of duetting. In such cases, male and female coordinate their singing so closely that even trained ornithologists are often deceived into thinking that only a single individual is holding forth. Not that all duetting birds are created equal; instead, they are made, built upon shared experience and the kind of careful listening and responding that characterize an experienced, world-class musical ensemble . . . or a long-lasting, mutually committed, domestic relationship. In the case of one notable duetting species, Australian magpie-larks, it was found that young couples are comparatively sloppy in their shared songs, leaving irregular gaps. With time and experience, and possibly the withstanding of stresses no less than the overcoming of obstacles, devoted couples become expert at anticipating the other's behavior, covering for mistakes, and, together, creating a whole (in this case, a defensible territory) that not only exceeds what either individual could achieve alone, but that also improves with time.[23] The title of the research article pretty much speaks for itself: "Temporal Coordination Signals Coalition Quality."

Here is yet another practical payoff to monogamy, a benefit of being in a "coordinated coalition." This one might seem as dry and unexciting as the simple accumulation of "stuff," à la beavers, but it is nonetheless real and genuinely rewarding in its own way.

Evolutionary geneticists have long been aware that sexual reproduction imposes substantial costs on any species, notably, the so-called "cost of meiosis." This is the unavoidable fact that each parent has only a 50 percent genetic connection to his or her offspring (compared to a potential 100 percent identity were that individual to reproduce asexually). Sexuality imposes other costs, too, including the time, energy, and risk spent trying to find a suitable mate and then persuading him or her to actually act on the mating impulse.

Among many animals, all that persuading—courtship—is costly. Individuals labor to primp and preen, engaging in seemingly bizarre and wasteful antics, all designed to attract a sexual partner. And human beings, of course, are no exception. To our knowledge, no one has yet measured the "mating effort" expended by male and female *Homo sapiens*, but it is doubtless immense, including beauty parlors, muscle-building gyms, tanning salons, painful diets, the purchase of fancy and/or enticing clothing, "hanging out" in bars and nightclubs, and attending parties and events that are often of little appeal in themselves, except for the prospect of maybe encountering someone of potential interest. These and a million other attempts to obtain a willing partner are literally expensive, however one measures it.

To be sure, monogamy isn't a cure-all for these costs. We would like to think that married couples continue to keep themselves healthy and attractive, and of course, some individuals—regardless of their marital status—expend time and energy prowling for additional partners. Nonetheless, once monogamous animals have located a suitable mate, they are able (glass half-full) or constrained (glass half-empty) to limit their extra-pair mating efforts, and the same applies to human beings. Time and energy not spent trying to attract, seduce, or otherwise engage with an extracurricular partner can be spent, instead, on . . . well, anything!

Chapter 7

LOVE

GIVEN THAT BIOLOGY IS ALL ABOUT GENES, reciprocity, material payoffs, "coalition quality," and other hard-eyed, unemotional, coldly calculated payoffs, where does *love* fit in? In the song "Some Enchanted Evening," the audience in the musical *South Pacific* is told that when it comes to explaining love, fools give reasons, whereas wise men don't even try. We're going to try.

To begin with, love is neither more nor less than a mechanism, a means to an end rather than an end in itself, a device that evolved by natural selection to facilitate breeding, the transfer of genes into the future. It is how nature gets people to obtain all those seemingly hard-eyed, unemotional, coldly calculated payoffs. That it feels good—very, very good—is testimony to how effective it is, not to mention the fact that, by and large, things we love are good for us. Imagine, for example, that some of our ancestors just *loved* eating arsenic, bashing their heads into rocks, juggling porcupines, or having sex with giraffes: They wouldn't likely have become our ancestors after all. Thus, love is not a "mere device," like the fictional Golux in James Thurber's novella, *The 13 Clocks*. It is, instead, a genuine thing of great potency. Like Golux, who admits to making things up, it is also unreliable. But love is nonetheless a potential source of wisdom and guidance, although—like our friend the Golux—it is imperfect.

You might love soft drinks, alcohol, or cigarettes, passions that almost certainly do more harm than good. Addiction is defined as behavior that is hurtful to oneself, but is engaged in, often repetitively, nonetheless. There is ample evidence that addictions themselves are devices that plants use to spread their genes around, via animals. For example, think of the success of sugarcane, coffee, coca, and tobacco. Addictive plants hijack humans to propagate their genes, and in general, these addictive plants are unlikely to promote the genes of their human victims. Alcoholics and narcotic and cocaine addicts promote the success of grapes, opium poppies, and coca, as well as the success of the narco-traffic dealers, typically at the expense of the downstream user. Addictions to plant derivatives in particular are neither new nor unique to Eurasian cultures. Our point is that "love" may be obsessive, demanding, compelling, and even at times self-destructive, rendering it experientially similar to being hijacked by plant substances.

Unlike addictive substances, however, love has at least the prospect of benefiting its "victim." But no guarantee. In its iconic form, of course, love is "for" another person. But really, it's for yourself. Being in love feels good, or at least compelling—to oneself. Natural selection has induced people to love others when doing so produces a situation by which their genes are likely to have benefited in the past. Love of one's children induces parents to be caretakers, often enduring costs, risks, impositions, or simple indignities that wouldn't be tolerated on behalf of someone else—which is to say, on behalf of genes not one's own. Love of children for their parents induces them to pay attention (and sometimes even offer respect) for adults who are likely to have the children's interests in mind, and whose age, experience, and power—combined with biologically infused benevolence—make them worth loving.

Then there is love of one adult for another. What's *that* all about? Lets look first at how it manifests itself.

Genuine love produces not only desire for the loved one, but for his or her well-being, coupled with a willingness to nurture the "love object." Yet the intense sensation of "falling in love" may well have a different structure and meaning than "staying in love" over time. *Falling* in love is often more narcissistic, with intense neediness and preoccupation, as well as sexual desire or erotic sensation, but not necessarily nurturance. More about this later, but for now think about the pernicious forms of love that result in stalking, where the "love" is more like a prey drive, the desire to capture, and contain control, or even kill the object, who is in fact a victim. Erotomanic stalkers feel love for their targets, but without knowledge, nurturance, or care for them. Their love is purely self-centered, a fixation, without any reciprocity or even genuine desire for the well-being of the other.

It is very important to remember, and we cannot say this often enough, that the word "love" is encumbered by too many meanings. It can mean benevolent interest and intent for another, or it can mean a compulsive fascination, obsession, and preoccupation, without any genuine desire for interaction or a positive outcome. When you hear someone profess his or her "love," ask which kind of love is meant: a deep yearning for another's well-being, or a masturbatory, narcissistic obsession.

Love can be "explained" by pathology, physical attraction, shared interests, the tweaking of particular brain regions by certain hormones, as well as the actions of specific neurons that have only recently been identified. (More on this later.) As to *why* love occurs among adults, evolutionary biologists such as ourselves point unblushingly to its selfish underpinnings, no less than to the love of parent for child and vice versa: When healthy, love is a naturally selected mechanism that makes it more likely that the lover will be biologically successful, not in self-gratifying isolation but because of his or her relationship with the beloved.

The miracle of monogamy typically includes many flavors of love at once or in sequence. A romance may begin with "chemistry," erotic desire plus intense preoccupation with the loved one. The lovers share time, space, and sex, organizing a joint lifestyle; mutual nurturance develops, and this may be enough. There are plenty of monogamous couples who stop with happy housekeeping, and do it for years. Others add child-rearing, and here the payoffs become both greater and more fraught, because "love" has to expand from amorous and sexy dates, and all the collaterals of intense mutual connection, to include those little intruders who require nurturant love, and only nurturant love, not the other varieties.

Children eventually become independent (especially in modern Western society), usually just as the monogamous couple are aging and becoming needy themselves. Ideally, the love the children received rebounds backwards, toward filial piety, with children taking care of parents.* At the same time, there emerges another biological advantage of monogamy for the original pair: The more self-sustaining and independent the parental "unit" can be, the more their children are freed to live their own lives without huge obligations toward those parents. And this, in turn, can increase the biological success of all concerned.

Let's spell this out a bit more clearly. Successful children should have enough resources and emotional bandwidth to help their parents in their old age, without serious decrements to the flourishing of the children or grandchildren. However, it is not uncommon—especially in America, whose social safety net is the least developed in the industrialized world—to see middle-aged "children" of the so-called sandwich generation balanced precariously between the demands of their sick, aging parents and the needs of their own offspring, with hardly a moment or dollar for themselves. Such circumstances, difficult enough to bear, are doubtless more challenging yet when borne alone.

As we've seen, reproduction looms large in all aspects of biology, and monogamy is no exception. It is neither exaggeration nor heedlessly pronatalist to note that whether or not people have children,

*Like love, filial piety is a cross-cultural universal, although more strictly defined and observed in some cultures (e.g., Korean) than others (e.g., Californian).

many of their inclinations were formed in a world in
offspring was the sine qua non of success. And so it
prise that even though a deeply loving couple m
whatsoever to become parents, much of their love i
on a biological foundation that largely owes its ex
that in the past it contributed to reproduction.

Love can assuredly take place without r
couples are every bit as loving as their heterosexu
it takes deliberate, sometimes heroic effort for
(turkey-baster babies, artificial insemination, in vit
only recently have these options become availabl
B. Toklas and Gertrude Stein: monogamous, lo
ductive. Elderly people can certainly fall in lov
despite being undeniably postreproductive. Re
romances are commonplace, especially when
sick or deceased. One of our grandfathers rem
love of his life passed away, and he remained
when his second wife died as well. (The old m

There are some who love each other so m
ing so much fun with their own shared living—
have children, so as to avoid diluting their mut
tions, and these are those who elect to be chil
the Earth's environment, or a pessimistic sens
duce children into what they perceive to be a
ing world. Whatever the specifics, there is no
reproduction is incompatible with love.

At the same time, reproduction can tak
the great majority of animal species, it does.
riages have occurred for millennia, doubtles
fertilization requires no particular sentim
carefully meted out and introducted to eac

Love, we must conclude, is neither
baby-making. Nor is it always in the biolog

beloved. There are examples of people sacrificing themselves for e, and even pathological cases in which someone "in love" does at harm to his or her beloved. In the vast majority of cases, however, is far more benevolent in its effect. Like gravity keeping planets rbit, love keeps people in positive relationships. But whereas gravin't designed to maintain astronomic propriety—it just *is*—love is igned" (by natural selection) as a way of enhancing the evolutionary ess of those involved. Putting it differently, to avoid any misunderding: Individuals and their genes that, as a result of experiencing thing called love, left more descendants, were those that predomd in the population. And so, evolution bequeathed us—who cone the outcome of all that generational descending—a capacity for indeed, a need for it, no less than our need for food or for sleep.*

Since love isn't limited to a sexual, romantic, bonding relationbetween adults, it is worthwhile to discuss the varieties of love, use false preconceptions often imperil those attempting gamy. For example, the intense obsessional hyperfocus of "being e" naturally diminishes over time. How do people gain satisfacnd pleasure from one another when acute romance fades?

The immediate onset of sexual love, or "falling in love," is like a rary gale, a brushfire, a compelling but short-lived manic e that psychologist Dorothy Tennov labeled "limerence."[24] by contrast, is an account of mature adult love, from Louis de res' novel *Corelli's Mandolin*. A wise, widowed father is telling ghter about the love that had united him and his wife.

> ove is a temporary madness, it erupts like volcanoes and then subles. And when it subsides you have to make a decision. You have work out whether your roots have so entwined together that it is conceivable that you should ever part. Because this is what love is.

who don't respond to their need for food or sleep would have left fewer descen- for love.

> Love is not breathlessness, it is not excitement, it is not the promulgation of promises of eternal passion, it is not the desire to mate every second minute of the day, it is not lying awake at night imagining that he is kissing every cranny of your body. . . . That is just being "in love," which any fool can do. Love itself is what is left over when being in love has burned away, and this is both an art and a fortunate accident. Your mother and I had it, we had roots that grew towards each other underground, and when all the pretty blossoms had fallen from our branches we found that we were one tree and not two.

Why be "one tree and not two"? Because when and if it works, the outcome is likely to be stronger, more resilient, less vulnerable, and more fulfilled emotionally, physically, intellectually, reproductively, economically, socially . . . in a word, biologically.

Chapter 8

ATTRACTION TO OTHERS

BIOLOGY AND MONOGAMY COEXIST in a complex and uneasy relationship. Considering "solutions" to humanity's nonmonogamous impulses, we've talked thus far about various biological payoffs to monogamy, each of them a substantial prod in favor of one-to-one mateship. Don't forget, however, that what makes human monogamy such a miracle is that to succeed, it must work against a great big biological obstacle: the altogether natural attraction that people feel toward other individuals.

According to Jackie Mason, "Eighty percent of married men cheat in America. The rest cheat in Europe." The reality, comic exaggerations aside, is that infidelity is frequent, everywhere. Furthermore, this reality exists because it is founded, at least in part, on genuine biological reality. This is delicate territory, however: As we have seen, many people confuse "is" with "ought," fearing that to describe and understand human nature is to justify people's worst inclinations, a kind of "Twinkie defense" that the genes justify the means.

Such justification is diminished, however, by the realization that any extracurricular sexual attraction is founded—like monogamy itself—on evolution, not on divine dictate or a DNA-based version of original sin. Moreover, sexual adventurism has its downsides: risk of disease, emotional injury, even loss of one's initial relationship, not to mention the almost certain costs that come with deception. At the same time, because philandering also threatens the philanderer's spouse, sexual jealousy is a widespread human emotion, and not only because an unfaithful individual may be prospecting for another relationship. If the "infidel" is a woman, her male partner is threatened with the prospect of rearing a different man's child, and, if a man, it presents his female partner with the possible diversion of time, energy, and money. So, just as it is part of human biology to be *tempted* by the prospect of additional sexual partners for oneself, it is also part of human biology to be *threatened* by sexual infidelity on the part of one's partner.

One pillar of successful monogamy, therefore, comes from acknowledging the temptation as well as the threat, with each partner recognizing that he or she is apt to be tempted on occasion, while also liable to be heartbroken when and if the other behaves similarly. The solution? Not easy, but straightforward and surprisingly effective, so long as it is *reciprocal*: a kind of mutual disarmament pact in which each participant agrees: "I will not screw around and drive you crazy, while you, in return, will not screw around and drive me crazy." As Johnny Cash famously sang, "Because of you, I walk the line." We call it "Mutually Assured Monogamy."

Reciprocity of this sort goes beyond that of the pygmy marmoset; it involves conscious, intentional determination to behave in a seemingly selfless manner, but one that is ultimately backed up by a biological underpinning of shared benefit.

Many animal species experience what biologists call "mate-enforced monogamy," in which male and female each "walk the line" because of the other's behavior. Sometimes the enforcement is aggressive, albeit comical, as in the case of dung beetles. Pairs of dung beetles labor long and hard, rolling up a ball of poop within which the female will lay her eggs. Males, if given the opportunity, will nonetheless court other females in addition to the one with whom he has just collaborated, so long as his mate is tethered nearby, unable to intervene. But if released, the female—almost visibly angry, even to a human observer—may then push him into the pile and roll *him* around for a bit.

When threats to the union arise, some animals move rapidly from cooperation to confrontation. The Peruvian warbling ant-bird is one of those duetting species discussed earlier. They are typically monogamous, with male and female entering into vocal as well as other forms of carefully orchestrated cooperation. When researchers experimentally played the recorded calls of a different pair of ant-birds, suggesting that these were intruders who might take over the residents' territory, ant-bird couples responded with their usual well-coordinated warbling song, indicating (and perhaps reinforcing) their united front. But then the biologists played the song of just a strange female, whereupon things were quite different: The male resident ant-bird warbled his little heart out, evidently trying to attract a new potential paramour. His mate, however, responded by interrupting the male's love-song with a rather loud and raucous call of her own, not otherwise heard, that essentially endeavored to jam the communication between would-be philandering male and his prospective enamorata.[25]

Peruvian warbling ant-birds; female left, male right. Photo by Joseph Tobias

In other species, mate-enforced monogamy may involve enhanced sexiness, as among European starlings. When researchers placed additional nest boxes within the territory of a mated pair in this species, the suddenly "wealthy" males responded by courting new girlfriends, who were evidently attracted by the choice real estate (good nest sites are typically rare in the starling world). Especially notable was the response of the original, mated females: They began sexually soliciting their "husbands," something not normally found among these animals, since the females in question had already laid their eggs. Perhaps they were seeking to enforce monogamy by distracting their otherwise errant mates.

In any event, human beings are certainly capable of enforcing at least a degree of monogamy upon each other, by using threats, promises, clearly stated expectations, and a firm knowledge of infidelity's likely consequences. Perhaps the best mechanism involves creating a benevolent disarmament pact with one's partner, which starts by identifying what behaviors are disallowed, along with the penalty for defecting. For example, if an affair is to be "punished" with murder, this hardly sets the stage for reasonable discussion or reconciliation—or for self-disclosure. Since cheating and deception are "natural," it may be hard to tell the truth when the consequences are very severe. But long-term monogamy requires openness and honesty, so the nature of the disarmament pact and its enforcement process and penalty system require discussion and mutual assent.

Mutually assured monogamy requires a kind of thoughtful dialogue about infidelity that is rare in any culture, but there is no reason that it could not become commonplace. After all, marriage itself is typically considered a contract, and mutually assured monogamy is exactly such an arrangement: Each signatory agrees "I won't have sex with other people, even if I may want to," while acknowledging severe consequences in the event of default. For example, the relationship may end. Perhaps property will be divided unequally, with the larger share going to the betrayed person. It could mean months, if not years, of dismal psychotherapy, coming to terms with every aspect of each person's character flaws. It could mean that the defector has to go to Sexaholics Anonymous (a 12-step program for sexual addicts) for some period of indefinite time, working through all the steps including the ones that clearly acknowledge error and ask for forgiveness. All of these outcomes are long, arduous, and expensive, but if infidelity happens and the couple wishes to recover from it, then a lot of time and effort will almost certainly be required.

Lest this seem dreary and painful, a dose of ill-tasting medicine, let's be clear that restraining one's biological appetite for sex and romance with someone new needn't lead to a loss of happiness. Quite the contrary.

Earlier we suggested a biological definition of love. Here, then, is one for happiness: a deeply pleasant feeling of joy and satisfaction that comes from recognizing that needs are being met, that things are going right. People are likely to feel happy when they have just had (or contemplate having) a good meal, good news, or some other gratifying circumstance (sex, too). Brains light up with "happy" as a signal that something is right, with "right" being "likely to contribute to success." And love is closely allied to happiness, with both legitimately connected to the satisfaction of immediate biological needs, which in turn reflect deep-seated evolutionary goals.

According to the Dalai Lama, it is a valid and natural human endeavor to pursue happiness, which is not the same as pleasure. Pleasure is a brief sensation; happiness is a durable, long-term experience. Monogamy is thus made to order for happiness, although not without its own unique costs. It requires forgoing certain pleasures—sexual variety and romantic autonomy, the joy of the hunt, periodic sexual enchantment, the stimulation of flirting and of occasional conquest—in return for happiness: the long-term benefit of sustained cooperation in child-rearing, homemaking, life-sharing.

At the same time, sexual excitement needn't be lost. Here, again, biology poses what seems to be a great obstacle, since, as we have already described, human beings—like so many other animals—have been outfitted with a tendency to respond to the lure of sexual novelty. As an ancient limerick puts it:

> Once plighted, no men would go whoring;
> They'd stay with the ones they adore,
> If women were half as alluring
> After the act as before.[26]

In his poem *Don Juan*, Lord Byron wondered, "how the devil is it that fresh features / Have such a charm for us poor human creatures?" Byron didn't know of the Coolidge Effect, or its evolutionary basis, but he was clearly in touch with the phenomenon. And in his dark novella *The Kreutzer Sonata*, Leo Tolstoy recounted a particularly pessimistic view of marital fidelity:

> In life this preference for one lover to the exclusion of all others lasts in rare cases several years, more often several months, or even weeks, days, hours. . . . Every man feels what you call love toward each pretty woman he sees . . . [And] even if it should be admitted that Menelaus had preferred Helen all his life, Helen would have preferred Paris.[27]

Three hundred years earlier, Shakespeare famously described Cleopatra's extraordinary sexual power as follows: "Age cannot wither her, nor custom stale her infinite variety." And yet, even the all-too-natural, biology-based fondness for erotic variety (even when less than infinite) can be benevolently manipulated in the service of monogamy. People wanting to make monogamy work can take advantage of novelty's appeal, in the service of their own pair bonds. The trick is to introduce novelty—rather, the *feeling* of novelty—into monogamy, all the while knowing it is still the one you love.

Of course, it doesn't hurt to point out the shared interests that underpin all of these conscious decisions, helping them to work. But at the same time, it is naïve in the extreme to think that rational argumentation, in itself, will allow *Homo sapiens* to stare down the barrel of hundreds of thousands of years of evolution and say, resolutely: "I simply won't be attracted to anyone else. I won't! I won't!" We just aren't *that* rational.

Yet the power of that which is irrational, automatic, and instinctive isn't monopolized by nonmonogamy. There are pro-bonding, promonogamous aspects of human biology, too, and they also operate largely outside the realm of conscious intent. The biological underpinnings of human behavior are deep and diverse, including many potential roots that can be consciously stimulated and encouraged, then allowed to proceed "on their own." Just as a skilled gardener can grow strong, healthy, beautiful plants by adroit use of sunlight, water, and nutrients, people can develop a monogamy "green thumb," skillfully employing their own biology on behalf of their pre-existing, innate capacities. A lovely flower resides within a rosebush—part of its biology—and needs only appropriate care to blossom. A monogamous relationship, too, can flourish if mindfully nurtured, especially by identifying the biology that dwells within every human being, celebrating it, and turning it toward monogamous ends.

People, in short, can use their logic and common sense to set mental processes in motion that can then generate their own emotional momentum.

It is to these hardwired processes that we now turn.

PRO-MONOGAMY HARDWARE

We have identified a quartet of mechanisms that support monogamy, together constituting what might be called the Four Horsemen of the Monogamist. First, "attachment theory," initially propounded by the British child psychiatrist John Bowlby. It suggests that human infants and their parents connect automatically and deeply from a very early age, serving needs that go beyond mere sustenance. Attachment theory was initially developed to explain parent-offspring love, but it seems likely to have implications for monogamy as well. It describes what biologists call a "proximate mechanism," that is to say, something that helps explain the *how* of behavior, rather than the deeper *why*. More recently, scientists have also identified three primary additional proximate mechanisms, all operating via neurobiology, by which social attachment seems to take place: mirror neurons, neuroplasticity, and certain hormones, especially the endogenous "love drugs," oxytocin and vasopressin.

---

* Every behavior requires an immediate "how" in order for it to occur; it requires an evolutionary "why" in order for the "how" to have been selected in the first place.

First, attachment theory. Bowlby italicized the crucial role of infant-mother attachment, not only for normal emotional development, but for basic mental and even physical health. (Incidentally, this work made Bowlby the most-cited mental health theoretician of the early 21st century, exceeding Freud, Jung, Piaget, or Skinner.) Bowlby began his research during World War II, studying the reactions of babies who were separated from their mothers due to the massive civilian evacuations in London. Although well cared for materially, these infants often went through predictable phases of mourning. Some rebounded quite well, whereas others fell into a deeply depressed state known as anaclitic depression or marasmus. They literally turned to the walls, and quit eating and drinking. Some of them died.

According to Bowlby, basic attachment starts with every infant's need for a secure connection to an adult caregiver, especially during the crucial period from birth to two years of age, and particularly during times of stress—but not only then. Bowlby emphasized that the absence of a responsive, sensitive adult is itself a major cause of stress, and that "separation anxiety" follows the loss of such an attachment figure.

Attachment theory fits well into basic ethology (the biological study of animal behavior) and evolutionary biology, in that it emphasizes the role of crucial biological needs in a species that is helpless and dependent at birth. Indeed, Bowlby was aware of animal ethology and he understood that primates other than humans require more than food to thrive; his work was perhaps the first intentional foray into "human ethology." His research was conducted before the revolution in biological theory that occurred in the late 1960s and 1970s called sociobiology, but it has nonetheless fared well over time and continues to inform much work in psychology and psychotherapy today.

In some ways, the test of an important new way of thinking is that it becomes taken for granted, whereupon it eventually becomes "obvious." (Upon reading *The Origin of Species*, Thomas Huxley is said to have muttered, "How stupid of me not to have thought of that.") Everyone knows that infants need to be cared for, and adults—if they are to be evolutionarily successful—need to care. Attachment clearly serves both these requirements, and is thus adaptive for all concerned. The adult-infant relationship is, of course, initially asymmetric, but over time the young child develops additional, more balanced attachments toward siblings and other extended family members, leading to peer relationships and, eventually, romantic and sexual attachment with other adults. Nor does attachment stop there, since most adults eventually enter into the role of caregiver, first to children and then, often, to elderly parents.

For our purposes, attachment theory suggests that the human behavioral system is predisposed to form exactly the kind of deep interpersonal connection that monogamy also requires, and, moreover, that an adult-adult bond may well be part of a healthy continuum involving attachment generally, from infancy to old age. It is also interesting to note that Bowlby initially used the term "monotropy" to describe what he saw as a child's bias toward attaching primarily to one caregiving adult figure. That term is now defunct, as developmental psychologists have become convinced that infants are not, shall we say, monogamous, in their early attachments.

The earlier psychoanalytic view, pre-Bowlby, had been that childhood "dependency" was a problem that needed to be outgrown; attachment theory, by contrast, emphasizes its healthy aspects, as expressed into adulthood. Bowlby's work also did much to upend the psychologically brutal policy whereby hospitals used to restrict parental visits during a child's hospitalization. At the time, however, he was severely criticized by the scholarly psychoanalytic community, since attachment theory also went against the prevailing ascendancy of "object relations theory," whereby infant behavior was thought to derive less from the realities of actual experience filtered through biological needs and more from the infant's internal fantasy life.

Attachment theory has been extended to include adult romantic relationships,[28] based on the notion that early patterns of attachment—including the consequences of inadequate or inconsistent caregiving, as well as of positive, affirmative experience—set the stage for subsequent behavior in adulthood. Four styles of adult attachment have been identified, each of them redolent with implication for monogamy: secure, anxious, dismissive, and fearful. This conceptual system can be depicted as a kind of game-theory matrix, of the sort we have already encountered:

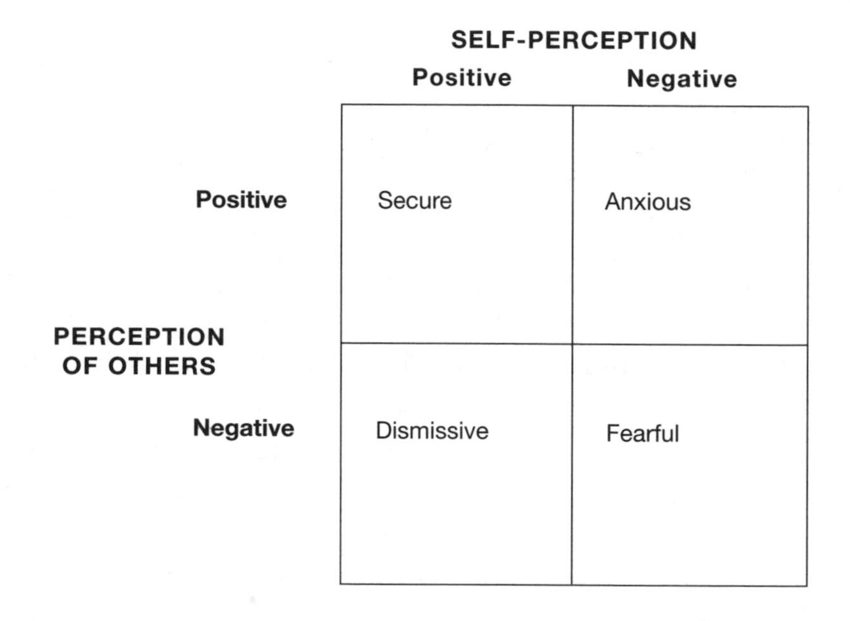

According to this view, people who grow up with a positive perception of themselves and of those to whom they were attached, especially when young, end up with a secure attachment style: seeking intimacy and also able to provide it. Those whose early experiences predisposed them to have a negative view of themselves but who perceive others positively are inclined toward an anxious attachment style, seeking intimacy but, because of low self-esteem, worried that others will be rejecting. A dismissive style, by contrast, is generated by experiencing a positive view of oneself but distrust of others; the result is a tendency to avoid close attachment, often claiming a lack of need for them. Finally, a negative view of self and of others generates fearful attachment—seeking intimacy but often reluctant to pursue it for fear that others will be hurtful (which, moreover, is what the fearful individual believes that he or she deserves).

The specific details of this model—including whether it is even valid—have been debated, and will doubtless continue to be controversial. The basic phenomenon, however, seems well established: Human beings are wired for attachment. Not only is it likely that one's early attachment experience—for good or ill—has repercussions when it comes to subsequent adult bonding, but attachment itself is a basic part of being human, at any age.

As Bowlby put it, "Attachment theory regards the propensity to make intimate emotional bonds to particular individuals as a basic component of human nature, already present in germinal form in the neonate and continuing through adult life into old age."[29]

Attachment theory is an important conceptual framework with likely relevance to monogamy, but it isn't a "thing." By contrast, our three other "monogamy support systems" have genuine physical existence, although since each is newly discovered, none is yet fully understood. Let's start with the single most exciting discovery of modern brain science: neuroplasticity.

Some species are born with a large dose of innate behavior, including built-in recognition of who to respond to, and in what manner. Human beings are not among them. We have a lot to learn. Yet, for many years, it was thought that nerve cells were incapable of new growth: You can't teach an old dog new tricks, and all that. Brains—especially adult brains—were assumed to be hardwired, incapable of significant growth or change. No longer. The dogma that brains, unlike, say, skin, couldn't repair themselves or establish new connections has been convincingly overthrown by what is perhaps the most stunning discovery by neurobiologists in the late 20th century.

Brains, we now know, are "plastic," which means that they are constantly molded, modified, and strengthened by experience. Specifically, dendritic spines (the microscopic outgrowths on the receiving ends of nerve cells) sprout in response to repeated stimulation, axons (the long cables by which neurons send outgoing signals) grow, synapses (the tiny spaces between sending and receiving cells) can be modified to facilitate, or inhibit, the transfer of information, just as genes within nerve cells are switched on and off in response to stimulation, and—even more striking for those trained in the earlier dogma of neural immutability—"neurogenesis" is real: New brain cells can and do grow.

To be fair, it was long known that the memory region of the brain—the hippocampus—was plastic, with new neurons produced here, even among adults. But otherwise, people were thought to be stuck with the brain cells they were born with—ruins us for those that died along the way. Not any more.

Not only can people (and, incidentally, old dogs) learn new tricks, even as adults, but as they do so, their brains literally change. For example, when serious professional musicians—who practice more than one hour per day—were compared with amateurs and nonmusicians, it was found that the brain regions concerned with fine motor control (the anterior superior parietal and inferior temporal areas) were largest in the highly trained group, intermediate in the amateurs, and smallest among the nonmusicians. Moreover, trained musicians appear to hear and read music with an area on the left side of the brain formerly thought to be simply a speech center. Amateurs who appreciate music listen with right-side brain regions, whereas professionals use both sides of their brains, but appear to specifically "read" music as a language, using brain regions otherwise devoted to language processing. Since it takes about ten years to become merely a good amateur musician, much less a pro, one can see that these so-called distributed brain functions hardly stop growing at age six. Although some little Mozarts are performing at four, the majority of very good musicians may not even start their studies until several years later. Clearly, the human brain changes and grows, and not just during childhood.

Consider this study that compared London taxi drivers with London bus drivers. Both have similar experiences navigating London traffic, and both jobs require good memory. But unlike the bus drivers, who essentially repeat the same routes from one day to the next, cabbies must vary their paths depending on the requirements of their passengers; in fact, in order to become a London cabdriver, it is necessary to pass a difficult test, requiring literally years of study, and reverentially known as "The Knowledge." Sure enough, compared to the bus drivers, cabdrivers were found to have a larger posterior hippocampus, the brain region concerned with acquiring and using complicated spatial information.[30] This is especially interesting because nobody becomes either a cabbie or a bus driver until after adolescence. Further confirmation: People who are bilingual develop a larger left inferior parietal cortex (a language-processing region) than do monolinguals.[31]

These findings are doubly intriguing. Not only do they reinforce the conclusion that brain regions respond physically to stimulation, but the cabdriver and bilingual speaker studies indicate that neuroplasticity isn't limited to brain regions that deal with actual body movements, or "motor functions," as shown by the musician research. Experiencing and thinking, too, can change the physiology and actual anatomy of the human brain.

Not surprisingly, most of the research associated with neuroplasticity has focused on its clinical applications, notably ways to facilitate recovery from strokes and other brain injuries.[32] For people who have suffered brain and spinal cord injury, it is especially exciting that connections between existing neurons can be altered—not just changes in receptor sensitivity, but also new growth on the part of existing neurons.

Most dramatic of all is neurogenesis, the actual birth of new nerve cells, demonstrated in at least certain brain regions: the hippocampus, olfactory bulb, and cerebellum. It remains to be seen whether neurogenesis occurs elsewhere in the nervous system, and whether stem cells can generate significant recovery of function.

There is, of course, considerable interest in the specific cell mechanisms—neurochemical, neurogenetic, neuroanatomic, neuroelectrical, and so forth—that generate plasticity. For our purposes, however, the precise details are less important than the bottom line: the basic fact of neuroplasticity itself, whose basic discovery is itself a revolution in brain science, and whose existence is no longer in doubt.

The precise relevance of all this to monogamy must be considered uncertain, although suggestive. When sensory nerves from various parts of the body are prevented from firing, cortical somatosensory representation for that body part is reduced; stimulation, on the other hand, increases their cortical representation. That is, stimulation causes the corresponding brain regions to be enhanced and enlarged.[33] Moreover, when a stimulus is reinforced (associated with positive sensations), the representation of that stimulus within the brain becomes larger yet, and even more responsive.[34]

The prospect beckons, therefore, that just as practice playing the guitar develops particular brain regions, practice interacting with one's beloved can develop brain regions that promote tolerance and mutual accommodation: in short, love and attachment. Plasticity also yields a biological counter to the tired clichés about "genetic determinism." Genes and their behavior-generating products, neurons, are not straitjackets but rather opportunity-providers that give people the tools to grow, develop, change, and adapt to new circumstances, including—we strongly suspect—interpersonal attachment. The ability to give up bad habits and to respond positively to good ones—not to mention good relationships—isn't just a pious, pie-in-the-sky hope, but a likely function of neurobiology. Thus, a biological take on monogamy builds upon neuroplasticity; rather than being constrained by genetics, the human capacity for monogamy seems likely to be empowered by it.

Next, mirror neurons.

As with neuroplasticity, mirror neurons were discovered only recently, and their significance is especially murky, but suggestive. It has yet to be investigated whether mirror neurons relate to bonding in general and to monogamy in particular—although we suspect they do. Here is the mirror neuron story, in brief.

Neurobiologist Giacomo Rizzolatti, at the University of Parma, in Italy, was leading a research team investigating the functioning of part of the prefrontal cortex of rhesus macaque monkeys. It had been known that certain motor neurons (responsible for generating movement) fired when the monkeys performed a particular action, such as grasping a banana. Quite by accident, however, Rizzolatti's group discovered that these same neurons would also fire when the monkey saw *someone else* grasp a banana. So these neurons didn't only control movement; they also responded when the animals perceived another individual engaged in the same movement. Moreover, it really had to be the *same* movement: A different array of neurons would fire, for example, when the monkey ate a banana, and similarly, if it saw someone else eating a banana.

These nerve cells, dubbed "mirror neurons," have been found in human beings, too. Although their existence is now beyond doubt, their function is less clear, although some exciting possibilities can be glimpsed. Consider empathy, for example, the experience whereby we can "feel someone else's pain," or when we imagine (or, better yet, know) what it would be like to "walk a mile" in someone's shoes. The prospect beckons that mirror neurons are the biological basis for "intersubjectivity," whereby people (and presumably certain animals, too) connect their own selfhood with that of others.

In one study, subjects viewed films of someone smelling a disgusting odor and then showing his distaste via predictable—and easily understood—facial expressions. By using fMRI (brain imaging) techniques, the researchers found that both being disgusted and seeing someone *else* be disgusted activated the same brain regions.[35] Follow-up research discovered that a similar process operates with regard to "tactile empathy"—the same brain regions light up when subjects are lightly touched on their leg as when they see someone else being stroked the same way.[36] A further step is the finding that when subjects receive a mildly painful stimulus (like a pinprick), cells that discharge in response also fire among subjects who are simply observing the event. This raises the question, as well, of how a masochist's or sadist's mirror neurons would respond to another person being poked (or worse). And it must be noted that schadenfreude—occasional, unanticipated pleasure at the discomfort of others—is real, suggesting perhaps that mirror neurons can have a darker side, even for the "normals" among us.

On a more positive note, neuroscientist V. S. Ramachandran refers to "empathy" or "Dalai Lama neurons." Could they also be "attachment neurons," "bonding neurons," even "monogamy neurons"? It certainly appears that human beings—along with many other species—are wired to respond to others in a manner that personalizes their experience as one's own: as a matter of literal, neurophysiological function. It isn't clear whether people are endowed with especially vibrant mirror neuron systems, in part because ethical restraints on human experimentation make it very difficult to conduct the necessary studies. But the intriguing possibility exists that the human capacity for language, and for complex sociocultural transmission, is something we owe to our mirror neurons.[37] Not so much "monkey see, monkey do," as "people see, people feel, understand, and copy."

As with neuroplasticity, mirror neurons also offer exciting clinical implications. There is some evidence, for example, that sufferers from autism are deficient in their mirror neuron system, which might help explain why autistic individuals are unable to "relate" to "normal" human communication.[38] And sociopaths—people who lack normal empathy for others—may be similarly afflicted, which would help explain why they can afflict others without feeling the guilt or the inhibitory pain characteristic of the rest of us, whose neurons typically fire uncomfortably in response to someone else's distress.

People adore experiencing things, not only directly, but also vicariously; hence the worldwide appeal of movies, theater, novels, sporting events, and computer games. It is one thing for mirror neurons to respond to the evidence of another's random action (seeing someone pick up a teacup), and just another step to feel someone else's pain (being poked with a pin) or, presumably, pleasure. And it is remarkable enough that at the neuronal level, such responses are quite specific to the experience in question. Is it unreasonable to go further and posit that these neurons not only mirror another person's experience but are also specific with regard to the individual whose experience is being mirrored? Mirror neurons fire, for example, when we see a stranger hit his thumb with a hammer; might they not fire even more readily, more persistently, or with more nerve cells being recruited if that someone else isn't a stranger, but someone already known? And who is better known than one's mate?

In Tennessee Williams's *A Streetcar Named Desire*, the pathetic Blanche DuBois famously notes as she is being led away to a mental hospital that she has "always relied on the kindness of strangers." Many of us—perhaps most—are moved by the distress of strangers, including that of the fictional Ms. DuBois. But aren't we even more moved by the distress of those we know and love? And if so, what about being moved by shared joys, accomplishments, experiences?

There is, accordingly, the possibility, maybe even a probability, that long-term relationships facilitate a greater degree of empathy at least in part because they set the stage for a greater mirror neuron response. If so, we predict that monogamous species (California mice, Malagasy giant jumping rats, pygmy marmosets, and so forth) have more mirror neurons, or more active ones, than do promiscuous or polygynous species. As for different individuals within the same species, one cannot help wondering whether monogamous individuals are endowed with more mirror neurons—or more active ones—than those who have multiple relationships. And if so, what is the cause and what the effect?

Perhaps having a more vigorous mirror neuron system leads to greater empathy, and thus enhanced affiliation—including love. The inverse could also be true, if—as seems likely—living together, and thus interacting on a regular basis, provides opportunities for any existing mirror neuron system to be activated. Neuroplasticity research has already shown that the more neurons are stimulated, the better they function and the more they grow: Nerve cells that fire together, wire together. So maybe monogamy and mirror neurons are mutually predisposed. The more people stay together, their neurons firing in response to the other's experiences, the more likely they may be to empathize with, feel for, understand, and, yes, love each other.

Never mind walking briefly—and metaphorically—in another's shoes. What about going alongside that person for literally decades?

Interestingly, there is also evidence (admittedly incomplete and sometimes contradictory) that species predisposed toward pair-bonded monogamy have larger brains than their polygamously or promiscuously inclined colleagues.[39] This may be due to the requirements of dealing with an "intimate other," the necessity of making frequent, prompt adjustments with regard to another adult. But it could also be argued that dealing with lots of others is no less stressful and equally demanding of brain power, which leaves the possibility that if monogamists are brainier, maybe this is because they are better endowed with the necessary mirror neurons.

Conversely, those same mirror neurons might also predispose people toward monogamy by making one less likely to inflict pain on another, with disinclination varying directly with the extent to which one can literally feel that pain. It is quite clear that "cheating," for example, inflicts emotional pain on one's partner. So, the more you can feel another's pain, perhaps the less likely you are to cheat.

In Shakespeare's *A Midsummer Night's Dream*, a love potion makes Titania, the queen of the fairies, fall in love with a commoner (who, for added effect, had previously been transformed into a donkey). Much hilarity ensues. A few centuries later, Wagner's *Tristan und Isolde* provides a more somber view of love potions and their powerful influence—on the human imagination if not directly on the libido. Indeed, the fantasy of an irresistible erotic stew has a history as ancient as it is compelling: Cupid's arrows were dipped in it, and anyone listening to popular music in the early 1960s will likely recall "Love Potion Number 9."

The truth, in this case, might be, if not stranger than fiction, accurately depicted by it. There just might really be a naturally occurring love potion. We present it as our fourth "Horseman."

Start with two species of voles (small meadow mice of the genus *Microtus*). Prairie voles are among those few monogamous mammals in which male and female bond together, forswearing other sexual partners, and in which the male vigorously guards "his" female. Their close relatives, the montane voles, are more traditionally mammalian: interested only in one-night stands. Enter oxytocin.

It's a prosocial hormone. When its natural release is blocked, mammalian mothers reject their offspring. On the other hand, virgin female rats infused with oxytocin fawn over another rat's offspring—something that doesn't normally happen. Female prairie voles (the monogamous species) require prior contact with a male in order to be sexually receptive. In one experiment, virgin females were pre-treated with oxytocin before being exposed to a male; they were sexually receptive right away, unlike controls, who received neutral saline and who rejected males' initial sexual advances.[40] Furthermore, oxytocin

Left, territorial defense in which a resident male prairie vole seeks to drive away a competitor.

Below, a mated pair of prairie voles, huddling with infant offspring in their nest.

PHOTOS BY LISA DAVIS AND LOWELL GETZ

A mating pair of prairie voles.

PHOTO BY LISA DAVIS AND LOWELL GETZ

doesn't only prime female prairie voles for sex, it is also key to their monogamous bonding. In fact, the sex isn't even necessary: Introduce oxytocin into the brain of a female prairie vole, even if she hasn't mated, and a celibate vole will proceed to act like Titania or Isolde, enthusiastically bonded to the nearest male. Moreover, if after copulation —when prairie voles typically establish their monogamous inclinations—the normal release of oxytocin is blocked, so is bonding.

The plot thickens. Oxytocin is also released in most mammals during labor and delivery, especially as the baby makes its way down the birth canal; it helps generate uterine muscle contractions. Indeed, the name oxytocin derives from the Greek for "rapid birth," and a synthetic form of oxytocin known as pitocin is often administered to pregnant women in order to speed labor. Oxytocin is also involved in stimulating the "let-down" reflex when a nursing mother's breasts physically release their milk. Interestingly, Chinese midwives have long known of a connection between nursing and childbirth, since tradition calls for applying ice to the nipples of women whose labor has stalled: Biologists now understand that stimulating the nipples leads to natural oxytoxin release (an effect that, incidentally, is not only more natural, but also less out-of-control than that produced by artificial pitocin "drips").

It probably isn't coincidental that the same naturally occurring hormone, oxytocin, has been recruited by evolution to facilitate both childbirth and milk release. Moreover, pair bonding is an important ingredient as well. It seems increasingly likely, in fact, that pair bonding in many species—including humans—involves variations on a deeper theme: the mechanisms initially deployed to achieve mother-infant bonding. And here is a case in which *Homo sapiens*—despite its capacity for thought (or, to some extent, because of it)—has special needs beyond those of other species.

Human beings are unusual in the helplessness of their infants, and thus, in the need for new mothers to bond to their newborns. And yet, people are also unusual in the pain that women experience during childbirth, which likely confronted our ancestors with an evolutionary conundrum: How to induce the victim of so much hurt to affiliate with the source of the anguish? People are blessed—sometimes, cursed—with a lot of intellect. We may be smart enough to acknowledge, rationally, the need for parental care-taking, but on the other hand, we are also intelligent enough to resent bestowing it on a tiny, odd-looking, squalling creature that has just caused so much pain. This dilemma might have been solved by employing the same hormone, oxytocin, that generates uterine contractions to also induce lactation when the nipples are stimulated, *and* to activate the same brain circuits that are involved in producing affiliation to another individual: the newborn infant.

The next step, and one that wouldn't have required much anatomic or biochemical innovation, is for this same hormone to be released during sexual intercourse in response to stimulation of the cervix, and especially during orgasm, as well as following erotic attention to female breasts. Which is precisely what happens. And which, in turn, contributes mightily to bonding—this time between mated adults.

Back to those voles, among whom oxytocin definitely is not the whole story. Montane voles don't bond, even when given artificial infusions of oxytocin. Female montane voles lack the brain receptors that are sensitive to this hormone. As for males, not surprisingly, their pair bonding is less reliably linked to underlying biochemistry, if only because males don't have to get over the potentially profound bonding obstacle of childbirth. But their brain circuitry is similar to that of females, involving vasopressin, a hormone closely related to oxytocin. The pair bonding (or lack thereof, in the case of the montane species) by male voles is predictable, depending on whether a particular genetic variant of a specific, identified vasopressin receptor gene is present. Montane voles don't have it, and they don't bond. Prairie voles do, and they do. Give oxytocin or vasopressin to female or male montane voles and nothing happens—they lack the brain receptors that mediate response to the hormones. So it's a matter of whether the appropriate receptors are present: The hormones oxytocin and vasopressin occur in every mammal tested. Presence or absence of receptors, on the other hand, is determined genetically, involving a gene that has been isolated and identified.[41] And this, in turn, correlates with sociosexual bonding—monogamy—or its absence.

This is only part of the story, although as far as we know, it is the truth, the vole truth, and nothing but the truth. But it isn't limited to voles. Those monogamous marmosets have more vasopressin chemically bound in their brains than do promiscuous rhesus monkeys. And when the prairie vole vasopressin receptor gene was inserted into normally promiscuous mice, the recipients became more socially attached to their mates.[42]

But what about *Homo sapiens*?

The human species, interestingly, has the same gene for oxytocin and vasopressin receptors that is found in voles and other animals. Moreover, oxytocin and vasopressin are released during sex in human beings, too—not just during childbirth and milk let-down. The actual attachment mechanism, in people no less than voles, appears to be connected to reward centers in the brain, by which oxytocin or vasopressin receptors stimulate the secretion of the neurotransmitter dopamine, which, in turn, feels good and induces voles (or people) to do more of whatever had generated the release in the first place. After a female prairie vole has mated, for example, researchers find a 50 percent increase in dopamine levels in her reward center.

In addition, people who describe themselves as being "madly in love" show particular activation in the same brain regions that are stimulated by cocaine. So love *can* be a kind of addiction. As for oxytocin and vasopressin, they also seem to be especially involved in identifying specific individuals, and thus connecting a rewarding bout of dopamine release with the person (or rodent) who helped evoke it. Among prairie voles, this recognition seems to involve olfactory cues in particular.[43] Moreover, oxytocin knock-out mice (genetically engineered animals who lack the capacity to produce oxytocin) also lose the ability to recognize conspecifics, even though their sense of smell remains unaffected.[44]

The role of oxytocin in identifying and bonding to specific individuals is not limited to prairie voles, nor to recognizing romantic partners; it extends to mother-infant attachment as well. Domestic sheep "imprint" onto the odors of their newborn lambs, rejecting strangers. This, too, is achieved with oxytocin, which acts via neurotransmitter activity in the ewe's olfactory bulb, priming her brain to latch onto her baby's odor.[45]

Does something similar happen in people? Human beings are less smell-driven than other mammals, but there is certainly much going on below the surface tht is unavailable to our conscious minds. And of course, there is every reason for sight, sound, touch, and a whole universe of other associations to combine and produce the end result, which often involves bonding—at least to some degree, and among some people.

This, in turn, leads to an important issue: variability. The simple fact that there are differences in the vasopressin receptor genes among voles leads to the suggestion that there is also variation in their faithfulness. The inverse of the question "Why do voles fall in love?"—with or without an accompanying rock and roll beat—is bound to be "Why is it that some don't?" And, once again, what about people?

Sure enough, variability in marital and romantic relationships in our own species appears to correlate with the presence of different forms (technically known as "alleles") of the receptor gene. Men with one notorious genetic variant, for example,* remain single at twice the frequency of other men, and, among those who do marry, are twice as likely to have serious marital difficulties.[46]

Nor is the potency of vasopressin, oxytocin, and their near relatives limited to sex and romance. A squirt of oxytocin, administered to human subjects via nasal spray, has been found to increase the recipients' altruistic inclinations, at least as measured by willingness to give money to another player in an experimental game.[47] Similarly, there is a correlation between the presence of a variant of the oxytocin receptor gene and various measures of trust.[48] Perhaps it is, in a sense, an empathy inducer. (It would be of more than passing interest to see if it activates the mirror neurons of whoever gets squirted.) Perhaps the day isn't far off when marriage counselors will be prescribing a dose of oxytocin or vasopressin spray.

But in the meanwhile, before such interventions are likely or possible, and before people begin worrying whether they might be desirable or ethical, whether they should be legal, or what they imply about the metaphysics of love, let's at least note that a material "cause" of empathy, attachment, affection, and, thus, monogamy, doesn't make these experiences any less real, or any less attainable.

Quite the opposite.

* The 5[1] flanking region of AVPR1A, allele RS3 334 – if you must know.

At this point, it should be clear that people have more than enough neural and hormonal infrastructure to support monogamy, evoked by evolution because of the many biological advantages (parenting, reciprocity, stuff, love, etc.) that pair bonding can provide. Our guess is that the four mechanistic pillars described in this chapter fit together something like this: Human beings have a profound need for attachment, beginning in infancy and continuing through adulthood. The benefits of attachment, however, aren't limited to childhood, including as they do meaningful payoffs to adults as well. Attachment itself (at any age) is encouraged by standard psychological processes, such as reward and reinforcement, and facilitated as well by mirror neurons, which, by promoting empathy, make for benevolent, prosocial, interpersonal connections. All the while, these connections are being literally structured by the brain's capacity for neural plasticity, in which nerve cells grow and brain regions develop in response to the continued interaction that defines attachment. And waiting in the wings, ready to provide an encouraging chemical environment, are those love-potion hormones oxytocin and vasopressin, along with their gene-based receptors.

The details, of course, are yet to be revealed, and will likely differ from our crude summary. It is 100 percent clear, however, that whether based upon Bowlby-esque attachment, mirror neurons, neural plasticity, or hormonal love potions—most likely all of the above, plus other factors not yet glimpsed—*Homo sapiens* is endowed with sufficient hormonal infrastructure, adequate neuronal hardware, and all the necessary behavioral blueprints and brain potential to support as much monogamy as anyone might desire.

Whether people *choose* it, on the other hand—and if so, how they proceed—is another matter, and one that depends a lot on conscious decision making and what the medical world calls "informed consent." That's next . . . and last.

Chapter 10

SOME TIPS FOR ASPIRING MONOGAMISTS

LONG-TERM MONOGAMY is neither necessary nor sufficient for a happy life. Nonetheless, those couples, gay or straight, who engage in successful, deliberate monogamy may enjoy a certain kind of bliss, and this chapter is directed to anyone hoping to join their ranks.

In this regard, Anne and Nat Barash, David's parents, are our inspiration and teachers. Nat died at 89 in 2005, and Anne at 90, in 2007. They had been married, monogamously so far as we know, for 64 years. It was not a first relationship for either of them. Nat had been something of a "man about town," and Anne was married briefly and unhappily before they met, in Montreal, Canada, when they were each about 25 years old. Both Anne and Nat were atheists and political radicals, and neither felt committed to sexual fidelity as a religious mandate or cultural norm. However, they dated, mated, and became inseparable, parted only by death. During that time, they spent perhaps half a dozen nights apart, and those for medical reasons. Not only were they inseparable, they remained happy, sexy, and loving.

What characterized their notable marriage was deep friendship, mutual respect, and constant engagement. They were bridge partners, dance partners, reproductive and housekeeping partners, and partners in their business, selling flowers at a small shop in the New

York City subway. The family joke was that Nat was in charge of the big decisions, such as who should be the Democratic nominee for president or how the United States should engage with China. Anne, by contrast, had the lesser role: determining where they lived, how they spent money, whether they should have children and how many, where they took vacations, what they ate, and where they got medical care. Nat took on Richard Nixon, then Ronald Reagan, and later, George W. Bush, never letting any of them off the hook. Anne decided when they should retire, whereupon they packed up and moved to Palm Springs. It was a perfect balance of power between best friends. Each trip to the hardware store was a date. Weekends were spent at yard sales. They had the best collection of used toaster ovens on the West Coast.

Lest this sound unrealistic, it should be noted that they also fought, but nonviolently. In their final years, Nat had an implanted defibrillator and Anne developed a neurogenic bladder. Despite these impediments, they flirted and were sexually active until the end. Since they were not virgins when they met, in the strictest construction they were serial, but successful, monogamists. Have no doubt: Monogamy can work!

There are in fact many ways to gain insight into successful monogamy, of which "case studies" such as Anne and Nat Barash are merely one example. Intriguing parallels also exist between evolutionary biology and economics, with biologists' "fitness" mapping closely onto economists' "utility."[49] And in fact, strategies of investing in the stock market make good analogies for understanding monogamy, in animals no less than people, with biological success paralleling "profit" in the world of economics. Some investors have made billions on short-term gambles, buying and selling quickly without given commitment to any single stock or concern about the fundamentals of a particular company. Being a day trader or hedge fund manager requires quick thinking, lack of emotional involvement in any given equity, and focus on one thing: making money. Theirs is a promiscuous strategy, incorporating portfolios with rapid turnover.

Need we point out, however, that not all such investors get rich, or remain that way? Moreover, some—like the infamous Bernie Madoff—are deeply unscrupulous and cause a great deal of harm and anguish. Others understand the benefits of honesty as well as the disadvantage of buying and selling too rapidly and impulsively, which is to say, the value of being in it "for the long haul." Warren Buffet became rich and famous by being careful with his investments, knowing the fundamentals, and not trading too quickly. His conservative style is more "monogamous," reminiscent of Anne and Nat.

To be sure, buy-and-hold investors also need to understand the danger of sticking too long with a losing portfolio, although, like divorce, disconnecting from a plummeting stock is often difficult. Hence, some investors install automatic stop-loss provisions on their holdings, ensuring that if a stock goes below a certain point, they get out rather than helplessly watch it nosedive. The worldwide recession that started in late 2007 is proof, were any needed, that there are bubbles in stocks and in real estate, as in relationships. Buy-and-hold is no longer considered a sacred golden rule. Better to sell an investment, even at a loss, than to lose more than you can afford. At the same time, the best guarantee of coming out ahead is making good investment decisions in the first place, with an eye for quality, which is to say, sound "fundamentals." Bubbles don't last—in relationships or in finance.

Just as smart investors evaluate the strengths and weaknesses of a potential acquisition, a would-be monogamist is well advised to undertake a personal assessment of his or her value as a potential mate, and also that of any prospective partner, in terms of the three basic biological "goods": good genes, good behavior, and good resources. It may sound cold and hard-eyed, but Malagasy giant jumping rats, dwarf fat-tailed lemurs, California mice, and so forth do just that (albeit, we presume, not intellectually), and there is no reason that human beings should settle for anything less.

Unfortunately, careful assessment of oneself and others is surprisingly difficult, especially for young people, which is why adolescent relationships are nearly always brief: The parties involved are of uncertain social, economic, and behavioral worth,* so what attracts young people is typically a volatile mix of mostly physical cues. Among *Diplozöon paradoxum*—those parasitic worms in which sexual partners meet as adolescents, after which their bodies literally join in permanent, monogamous union—there are no expectations of reciprocity, orchestrated cooperation, burden-sharing, or love. But *Homo sapiens* expect more of their mates; hence, they are unlikely—and certainly ill-advised—to transform an uninformed, early life crush into a permanent commitment.

As a general rule, it is also a good idea to take seriously the biological pattern known as "assortative mating," whereby individuals are likely to pair with others who are roughly similar, in intelligence, education, social standing, and even physical traits such as height, weight, and attractiveness. For quick confirmation, just peruse the weddings section of any newspaper: Couples consistently mirror one another, something true among animals as well. Opposites attract? Yes, when it comes to basic heterosexuality. Otherwise, similarity is the rule.

In the case of Anne and Nat, neither went to college and neither came from a wealthy family. Their wedding would not have been noted in the *Montreal Star* because they were working-class. However, both came with little negative baggage and were of roughly similar social capital. Both were attractive, although not gorgeous. Both were very smart, hardworking, and from families in which they were loved and valued, and so each had high basic self-esteem and good mental

*This is not a matter of legal or ethical worth, but something more fundamental, biological, and perhaps troubling.

and physical health. Neither had problems with addictions, compulsions, anxiety, or mood swings. They both enjoyed dancing, politics, bridge, and, most of all, each other. But even before Anne and Nat met, they were healthy and well grounded in mature, sensible relationships with their families and the world.

The exceptions to assortative mating are primarily cases in which men marry down socially because they are attracted to youth, beauty, and raw sexual energy, and women marry up ("hypergamy"), mostly for money. It is not uncommon for wealthy men to marry younger, beautiful women whose primary appeal is, frankly, being young and beautiful. By contrast, it is relatively rare for successful women to marry down, so called "hypogamy." Some notable women use and discard "boy toys" much as rich men use and discard lovely young women, but monogamy is especially unlikely to persist when the woman is of substantially higher biological and social "worth" than the man. (This harkens back, we suspect, to *Homo sapiens*' basically polygynous nature.)

You might want to take a lesson, at this point, from the pied flycatcher. This small forest-dwelling bird is reputedly monogamous, but in fact, already-mated males will sometimes court an additional, "secondary" female—which is to say, a mistress—with whom he mates, but who then must rear the offspring by herself, since all of the male's parental efforts are expended with his primary mate.[50] The secondary female is therefore duped into becoming a single parent, and she inevitably rears fewer offspring than she would have if she were the male's primary mate. For their part, males achieve this deception by courting secondary females at some geographic distance from their primary female and nest site, not unlike the cad who removes his wedding ring while wooing a girlfriend on the side—often in a different city. But secondary females aren't without counterstrategies: In particular, by prolonging courtship and delaying reproductive commitment, they increase the likelihood that a male's simultaneous romantic entanglements, if any, will become apparent.

The case of the pied flycatcher does not necessarily argue for sexual abstinence, especially given that birth control (not available to pied flycatchers, whether primary or secondary) is an option for human beings. But it does suggest a certain wisdom in postponing long-term commitment, and particularly reproduction, until each partner has had adequate time to plumb the depths of the other's commitments and character.

Temperament also matters. It is increasingly well established that human beings, like many other creatures, are born with certain emotional styles that change remarkably little over time. Now-classic studies by researchers Stella Chess and Alexander Thomas identified three basic styles in very young babies, showing that these styles were recognizable by six months of age and stable over the course of years.[51] These distinct temperaments include "easy" babies, who eat, sleep, and socialize with equanimity; "difficult" babies, who are prone to colic, irregular sleep, and cranky social interactions; and those who are "slow to warm up," shy with strangers but ultimately tractable and consolable. Those who showed early social and stranger anxiety grew up to be somewhat more neurotic than the other groups.* Irritable, cranky babies became irritable and cranky adults, less easily consoled by others, less attached, more self-centered. And the easy babies who possessed natural abilities to self-soothe and calm down were less prone to emotional illness than the other two groups.

Animals clearly show temperament traits that are variable across species and consistent within individuals. For example, experienced dog breeders can predict the ultimate temperament of very young puppies by subjecting them to a simple series of tests. Puppies showing a lot of submissive urination under stress are more likely to be anxious as adults—with separation anxiety, fear biting, and other worrisome traits—than puppies who submit to pressure and bounce back

*The word "neurotic" has fallen out of favor, along with other concepts from psychoanalysis, and yet we don't think a better term has been coined to describe individuals who are reactive, emotional, and at times irrational.

quickly. We are fairly confident in knowing the temperamental traits of our horses, cats, and dogs, from the time they are quite young, because these attributes don't change much over time and development. Feral cats nearly always stay wild, although they can make friends. Easily spooked horses have panic attacks, although they can be trained to overcome their fears. "Hard" dogs have dominant temperaments and make great police or working animals, but they may not be ideal family pets.

Accordingly, we advise assessing one's own temperament and that of any prospective mate, looking at reactivity, social inclinations, dominance patterns, and mood, because people, too, have character and temperament, and for an aspiring monogamist, looking for honest character and stable and resilient temperament in mates is much more important than facial symmetry or a waist/hip ratio of 0.7, the universal ideal of female beauty.[52]

Of course, traits that are "good" for one person may be deeply troublesome for someone else, and vice versa. Here are some interesting results from research on a mostly monogamous mammal, the old-field mouse, *Peromyscus polionotus*. It is not quite as faithful as the California mouse we encountered earlier, but impressively monogamous nonetheless. In one study, male old-field mice were given a choice between two virgin females, and their preferences were recorded. Then these males were paired, in some cases with the preferred lady mouse, and in others with the disfavored one. Interestingly, more surviving pups were produced when males were allowed to mate with their first choice than when forced to breed with the rejected female.[53] Evidently, male old-field mice know something about who is a "good" reproductive partner.

But the situation is more interesting yet, because the research also showed that mating preference in this species isn't simply a matter of males choosing healthier females. In a second part of this experiment, female mice were divided into two groups: those preferred by a given male and those who were rejected. Different males—who had never encountered females from either group—were then paired either with the preferred or the rejected females. In this case, there was no difference in the number of offspring produced, which suggests that the rejected females weren't inherently inferior to the preferred ones, just that they were less suitable as reproductive partners *for the males who rejected them*. In other words, and at least in this species, a "bad" partner for one individual can be a perfectly "good partner" for another.

Can these findings be extrapolated to human beings? Probably. Although certain traits, such as personality disorders doubtless augur poorly for long-term success, others are probably specific to each potential partnership. One person's poison could be another's nutrition. Someone might be repelled, for example, by the prospect of sharing his or her life with a vigorous assertive personality, whereas a different person might revel in such a partner. To our knowledge, no one has yet studied the personality traits—psychological or physical—that characterize successful monogamous unions in the animal world. This would seem a worthwhile research agenda in itself, not to mention its possible relevance to *Homo sapiens*.

According to anthropologist Helen Fisher, there are three distinct mechanisms—lust, romantic love, and long-term attachment—which mediate, in turn, mating, pair bonding, and parenting.[54] Lust need not lead to romantic love, which in turn does not necessarily parlay into long-term attachment. Proceeding through this triad, each is likely to be calmer, more lasting, and also (not coincidentally) more comfortable and comforting than the one before.

Anyone seeking monogamy should be prepared to look beyond the exciting early stages toward the latter ones. It is especially important to find a good cooperator; in other words, a friend no less than a lover. This is not as easy as it may seem, although like so many things, it is facilitated by identifying what really matters. Women may be drawn to nasty, brutish men who seem sexy because they are strong, powerful, rich, etc. Such individuals may be fun in the short term but are unlikely to settle down to happy housekeeping, and disappointment inevitably awaits the woman or man who expects a major change in a spouse's behavior after marriage. Female lust and the propensity to fall in love may be as likely triggered by sexy sociopaths as by friendly gentle cooperators, partly because (in the unconscious whispers of their victims' DNA) those sexy SOBs may leave sons who are equally reproductively successful. This is one of many evolutionary traps for the would-be monogamist. Marrying a big shot or a gangster is about as likely to produce long-term monogamy as eating chicken would enable you to fly.

Men have a parallel problem. A woman who seems especially sexy may be just that, with little else to offer, and perhaps unlikely to apportion herself to just one guy. Meanwhile, a young virginal female is an unknown quantity in terms of her ultimate disposition. A woman who appears chaste and inexperienced may be that—or socially anxious and stubbornly neurotic. For a man seeking a good long-term monogamous cooperator, a vamp may be exciting, or a virgin enticing, but by using informed biology, men would be advised to keep looking until they find a woman who shares their values and cooperates well. The likelihood is that when it comes to choosing a potential mate, pygmy marmosets are far more interested in these qualities than in who has the most primate "bling."

Try viewing a prospective mate as just that: a *mate*, as in shipmate, no less than someone with whom to mate. A good friend, someone to share a small boat with in occasional rough seas, and who is also your lover, is more likely to make the successful move to monogamy than a hot bedmate with no additional social capital, who is liable to jump overboard, scuttle the ship, or not be available when the call goes out "All hands on deck." We therefore suggest falling in "deep like" first: a sincere and effective friendship. *Then* let the sexy stuff happen: Good friends make good monogamists (and better lovers, too!). And what is it that makes for good friends? Fair, consistent, and thus satisfying cooperation: reciprocal altruism, in the language of evolutionary biology.

When choosing a partner, whether first mate or last, don't be shy about taking advantage of the uniquely human attribute of language and complex communication. Human beings are remarkable in their ability to talk, and in our opinion they should do so a lot, especially if they are planning to attempt a life together, not least because (as we have tried to show) such a shared life is not only potentially rewarding, but (as we have also tried to show) fraught with biological pitfalls.

Earlier, we proposed mutually assured monogamy, and indeed, our recommendation resembles an arms control treaty between countries—albeit allies rather than Cold War opponents. Couples are usually uneasy with any open discussion of adultery, almost as if ignoring the topic will make it go away. We suggest the exact opposite: Just as alcoholism appears to be more frequent among people who grow up in families in which drinking is a forbidden pleasure, infidelity is more likely when it is literally unmentionable. Shoving things under the rug only makes for a lumpy rug. Our advice, therefore: Be honest and aboveboard. Everyone is tempted to cheat once in a while,* and there are circumstances that make it more or less serious.

For example, there was a decent man whose wife developed severe mental illnesses. Depression and psychosomatic complaints took up her entire life. She became an invalid. He did not leave her; rather, he paid her bills and scrupulously arranged and chauffeured her to her many doctor's appointments. He was kind, helpful, and solicitous, and every Saturday night he went to "a poker game with the boys" . . . in reality, to visit his mistress. The infidelity enabled him to be loyal to his wife, not sexually, but in every other way.

A contract for mutually assured monogamy should ideally take into account other potential vicissitudes of living together for decades; for

* Although doing so is quite another matter.

example, possible accident, illness, financial expectations and contingencies, how to deal with stepchildren, aging or demanding in-laws, etc.

Finally, reproductive discussions. We have noted that monogamy among animals is closely and complexly tied to the rearing of successful children, but also that human beings are unique among living things in being able to decide, consciously, against reproducing. What if one partner wants children and the other doesn't? Or if one or the other is infertile? Would it be a departure from monogamy if a couple in which the man had poor sperm quality allowed or encouraged his partner to have IVF with sperm from a friend or anonymous donor? Genetically, that would be the exact equivalent of cuckoldry, but socially, most people think of it as quite different.

The discussion points in mutually assured monogamy are long, detailed, and perhaps boring and beside the point when the moon is full and spring is in the air, but monogamy, if it is to last, must be for all seasons.

The various monogamy-relevant contingencies we've just described are unpleasantly like the fine print in a credit card agreement, something upon which nobody really wants to dwell. But those who have done so, who have dutifully considered these and other irksome contingencies, and have also reflected on the birds and the bees, the California mice and the Malagasy giant jumping rats, the beavers and the pygmy marmosets, those who have come so far in friendship and erotic attachment, making wise plans and decisions . . . even these admirable people sometimes make mistakes, and any good contract must consider that possibility. Then what?

Some reactions to infidelity involve considerable violence, physical as well as verbal. They are the equivalent of the nuclear option and may be, paradoxically, as "natural" as monogamy is "unnatural." Indeed, sexual betrayal or even the suspicion of infidelity is one of the major causes of murder cross-culturally. Although dire threats of mayhem may deter some adultery, it probably causes more of it to simply go underground.

So, what's the problem with having a few secrets between lovers, who are, after all, consenting adults? Is it fair to expect complete transparency and self disclosure? What is the problem with a "don't ask, don't tell" policy, or "even if asked, don't tell," or maybe a relationship in which it is assumed that there are a few secrets, but no worries so long as they don't become public? In theory, there needn't be a problem here (recall the Saturday night "poker game"). Perhaps an active, working, respectful, even loving relationship characterized by a few secret sexual departures on the side needn't be contraindicated. Perhaps. Except that then it isn't monogamy, but rather some other thing, more sexually diverse, and also, we strongly suspect, not likely to last, if only because paradoxically, just as human biology endows every healthy person with an occasional yearning for a new sexual partner, it has also bequeathed *Homo sapiens* with a deep discomfort if one's mate engages in the same behavior!

Furthermore, secrets are themselves a problem, if only because a relationship founded on deception is necessarily liable to be seriously flawed, or at least to involve less genuine relating. In addition, secrets have a habit of leaking out, and when they do, beware the ravages of jealousy—what Shakespeare, in *Othello*, called "the green-eyed monster, which doth mock the meat it feeds on." It may seem trivial or even passé in the oh-so-modern 21st century, but sexual jealousy is one of those biological atavisms, like peanut allergies, that can be lethal.

So the problem becomes how to create a contract for monogamy that attempts to prevent the possibility of sexual departures, while also allowing for some kind of repair and forgiveness process if "nature takes its course" and a sexual escapade occurs.

For some, the prospect of long-term marriage counseling, psychotherapy, and self-reflection is punishment enough. Achieving forgiveness in a marital breach requires endless apologies and replaying of who did what to whom, where, when, why, and so forth. It is not just a matter of "oops, sorry, forgot myself." We are reminded of the scene in *The Hitchhiker's Guide to the Galaxy* in which Ford Prefect and Arthur Dent are caught hitchhiking in an alien spaceship. The punishment: listening to Vogon poetry, the most tedious experience in the galaxy.

Marital repair can be a little like a Vogon poetry recital. Like it or not, it commonly takes years to achieve forgiveness, understanding, and reconciliation in the aftermath of marital infidelity. At least one nonlethal deterrent to adultery is the prospect of endless self-recrimination and reflection, with professional assistance only making the whole process more expensive, and occasionally more effective.

In some animals, such as beavers, the typical activity that keeps couples together is raising the family, offspring, grand-offspring, and other individuals who share some genes. However, as we have seen, social monogamy (housekeeping with one individual over the life span) is not the same as sexual monogamy (having sex with only one partner), since infidelity has been detected in virtually every species ever tested via DNA fingerprinting.

It remains true, however, that monogamy is achievable for many species, including our own, and that mutual cooperation and investment are the best predictors of sexual as well as social monogamy. Returning once more to David's parents, Anne and Nat, they were devoted to raising their two boys, while also providing considerable caregiving to other relatives, including their own parents, siblings, cousins, nieces, and nephews, within their large extended family. Their energy investment was reciprocal: Anne spent considerable time with Nat's sister and brother-in-law (even though she shared no genes with them), while Nat interacted substantially and benevolently with Anne's numerous nieces and nephews. When David was young, he wasn't sure which of his relatives were from his mother's side and which were from his father's. Everyone was treated pretty much equally, with a huge amount of time, energy, and money invested in the entire extended family. Anne and Nat's partnership was a full-time occupation, and also balanced as well as reciprocal.

Almost certainly, it helped that they were peers, comparable in attractiveness, intelligence, health, wealth, generosity/selfishness, and emotional stability. As it happened, they were also from similar ethnic backgrounds—although the overwhelming probability is that this was incidental and not in any way a prerequisite for success. There seems to be no reason for mixed-race couples, or those following different religions, for example, to have particular worries about their likely marital longevity. On the other hand, we cannot help being less than sanguine about proposed unions between individuals who don't "match up" when it comes to those characteristics that really matter—not skin color, religion, or political preferences, but basic, biological traits and compatibility.

Key ingredients, then, in this particular recipe for monogamy: sexual fidelity, kin selection, shared tasks (perhaps including child-rearing, but not necessarily), genuine friendship, physical/sexual/emotional compatibility, and serious reciprocal altruism, along with mutual cooperative inclinations and interests. Also, not least, a conscious commitment to make the whole thing work.

Finally, aging. The Beatles' Paul McCartney famously asked about still being needed, and still being fed, at 64. This simple song, with its basic dichotomous question, encapsulates an enormous amount: from taking care to being cared for. Well, Sir Paul has now passed 64 years of age, and he seems to be going strong. Perhaps 84 is the new 64? But this merely postpones the issue. The question is: What happens to monogamy when bodies get old and frail, illness happens, and there may be a very real loss in sexual togetherness and social investment, as the world closes in? Aging awaits everyone, but not at the same rate. Will monogamy tolerate an asymmetrical investment, if one partner is relatively hale, healthy, and even horny, and the other is frail, ill, or needy? We would hope that, having come so far, the answer will be found in simple loyalty and persisting love, but in practice, we have seen amazingly cruel departures from monogamy, just when one would have thought that such issues had been successfully resolved. Example: An attorney ran off with a very young woman, while his wife of 40 years was being treated for stage 4 breast cancer. This man simply could not stand the idea of illness and death, and rather than seeing his wife jaundiced and bedridden, and having to say goodbye, he decamped, choosing sexual diversion over his dying former partner.

In theory, the benefits of sexual and social monogamy persist—and maybe even increase—after many years together, when estate-building and child-rearing are over and it is time to simply take care of one another. Many people suffer after 65 from "benign senile forgetfulness," not dementia of the classic kind but a more subtle loss of memory for words, names, things, and locations. Two people together form a kind of social GPS system, orienting one another toward lost objects, plans, and living in the real world. It takes both David and Judith together to put in a new toilet apparatus, or to clip dog nails and trim goat feet.* Two cooperators can live more cheaply than two selfish individuals. Similarly, two physically diminishing people may not get as much done as they used to, but they can certainly accomplish more than if both were alone.

We close with our favorite image of monogamy, from Tomi Ungerer's children's book *I Am Papa Snap and These Are My Favorite No Such Stories*. We put this picture on our wall thirty years ago, and have stared at it, with increasing appreciation, ever since. Mr. and Mrs. Limpid represent the end stage of monogamy, not very sexy, but very important.

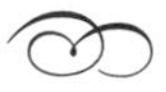

* Maybe this says something about our own incompetence, but that's the point: No one is omnicompetent, which is one of the things that makes partnerships so helpful.

From: Tomi Ungerer, *I am Papa Snap and These are My No-Such Stories.*
COPYRIGHT 1973 DIOGENES VERLAG AG ZURICH

Mr. and Mrs. Limpid are cooperators par excellence. Monogamy takes empathy—using those mirror neurons, acting upon one's attachment, intuiting what makes one's partner happy—as well as good intent. It requires a conscious decision not to fight about small things when bigger ones are at stake. It involves having each other's backs and compensating for each other's faults and weaknesses. It takes the strength, endurance, willpower, and intelligence of two beings to create a corporation whose sum is greater than either one alone.

The endearing image of the Limpids may be enough to inspire some to foreclose sexual opportunity in order to achieve interpersonal stability. Or maybe not. There are other ways to age, including organized retirement communities of people with similar interests who are neither paired up nor want to be. We know of retirement communities for professors, equestrians, and birdwatchers. Monogamy is just one way to live, and we make no claims of it being inherently superior to anything else. However, this book is about the sweet option of monogamy, for those intrigued by its potential and willing to pay its opportunity costs, as well as undergoing a mostly benevolent but nonetheless persistent chafing against someone different and yet attached, a semi–Siamese twin with an individual brain.

Human monogamy is something of a natural miracle, like the Grand Canyon, the product of years of exposure and the erosion—or at least, adjustment—of a separate self. Not good for everyone, but quite wonderful upon occasion, especially at sunrise and sunset.

Notes

1. D. P. Barash and J. E. Lipton. 2002. *The Myth of Monogamy: Fidelity and Infidelity in Animals and People*. Henry Holt/Times Books: New York.

2. M. Black. 1964. "The gap between 'is' and 'should.'" *Philosophical Review*, *73 (2)*: 165–181.

3. D. Lack. 1972. *Ecological Studies of Breeding in Birds*. Chapman & Hall: London.

4. K. Kraaijeveld, P. J. Carew, T. Billing, G. J. Adcock, and R. A Mulder. 2004. "Extra-pair paternity does not result in differential sexual selection in the mutually ornamented black swan (*Cygnus atratus*)." *Molecular Ecology 13* (6): 1625–1633.

5. O. Bray, J. Kennelly, and J. Guarlino. 1975. "Fertility of eggs produced on territories of vasectomized red-winged blackbirds." *Wilson Bulletin 87*: 187–195.

6. U. H. Reichard. 1995. "Extra-pair copulations in a monogamous gibbon (*Hylobates lar*)." *Ethology 100*: 99–112.

7. M. Wiles. 1968. "The occurrence of *Diplozöon paradoxum* Nordmann, 1832 (Trematoda: Mongenea) in certain waters of northern England and its distribution on the gills of certain *Cyprinidae*." *Parasitology* 58 (1): 61–70.

8. G. P. Murdoch. 1949. *Social Structures*. Macmillan: London.

9. D. P. Barash and J. E. Lipton. 2009. *How Women Got Their Curves and Other Just-So Stories*. Columbia University Press: New York.

10. J. Treas and D. Giesen. 2000. "Sexual infidelity among married and cohabiting Americans." *Journal of Marriage and the Family* 62: 48–60.

11. D. O. Ribble. 2003. "The evolution of social and reproductive monogamy in *Peromyscus*: Evidence from *Peromyscus californicus* (the California mouse). In *Monogamy: Mating Strategies and Partnerships in Birds, Humans and Other Mammals* (U. H. Reichard and C. Boesch, eds.). Cambridge University Press: United Kingdom.

12. D. J. Gubernick, S. L. Wright, and R. E. Brown. 1993. "The significance of father's presence for offspring survival in the monogamous California mouse, *Peromyscus californicus*." *Animal Behaviour* 46: 539–546.

13. D. Canton and R. E. Brown. 1997. "Paternal investment and reproductive success in the California mouse, *Peromyscus californicus*." *Animal Behaviour* 54: 377–386.

14. For a beginner's introduction to game theory, see D. P. B. 2003. *The Survival Game: How Game Theory Explains the Biology of Cooperation and Competition*. Henry Holt/Times Books: New York.

15. S. Sommer and S. H. Tlichy. 1999. "Major histocompatibility complex (MHC) class II polymorphism and paternity in the monogamous *Hypogeomys antimena*, the endangered, largest endemic Malagasy rodent." *Molecular Ecology* 8: 1259–1272.

16. A. Rylands. 1999. "Habitat and the evolution of social and reproductive behavior in Callitrichidae." *American Journal of Primatology* 38: 5–18.

17. A. W. Golidsen. 2003. "Social monogamy and its variations in Callitrichids: Do these relate to the costs of infant care?" In *Monogamy: Mating Strategies and Partnerships in Birds, Humans and Other Mammals* (U. H. Reichard and C. Boesch, eds.). Cambridge University Press: New York.

18. D. P. Barash. 2001. *Revolutionary Biology: The New, Gene-centered View of Life*. Transaction Publishers: New Brunswick, New Jersey.

19. R. L. Trivers. 1971. "The evolution of reciprocal altruism." *Quarterly Review of Biology* 46: 35–57.

20. A. Einstein and L. Infeld. 1938. *The Evolution of Physics*. New York: Simon and Schuster.

21. K. Sullivan. 1985. "Selective alarm-calling by downy woodpeckers in mixed-species flocks." *Auk 102*: 184–187.

22. Z. Liu, T. A. Nelson, C. K. Nielsen, and C. K. Bloomquist. 2008. "Microsatellite analysis of mating and kinship in beavers (*Castor canadensis*)." *Journal of Mammalogy* 89 (3): 575–581.

23. M. Hall and R. Magrath. 2007. "Temporal coordination signals coalition quality." *Current Biology* 17: 406–407.

24. D. Tennov. 1999. *Love and Limerence: The Experience of Being in Love*. Natonal Book Network: Lanham, Maryland.

25. J. A. Tobias and N. Seddon. 2009. "Signal jamming mediates sexual conflict in a duetting Bird." *Current Biology*: www.cell.com/current-biology /abstract/S0960-9822(09)00746-5.

26. L. Untermeyer, ed. 1956. *A Treasury of Ribaldry*. Hanover House: Garden City, New York.

27. L. Tolstoy. *The Kreutzer Sonata*. http://etext.lib.virginia.edu/toc /modeng/public/TolKreu.html, Chapter 2.

28. C. Hazan and P. R. Shaver. 1987. "Romantic love conceptualized as an attachment process." *Journal of Personality and Social Psychology* 52 (3): 511–524. C. Hazan and P. R. Shaver. 1994. "Attachment as an organisational framework for research on close relationships." *Psychological Inquiry* 5: 1–22. K. Bartholomew and L. M. Horowitz. 1991. "Attachment styles among young adults: A test of a four-category model." *Journal of Personality and Social Psychology* 61 (2): 226–244.

29. J. Bowlby. 1988. *A Secure Base*. Basic Books: New York.

30. E. A. Maguire, K. Woollett, and H. Spiers. 2006. "London taxi drivers and bus drivers: A structural MRI and neuropsychological analysis." *Hippocampus* 16: 1091–1101.

31. A. Mechelli, K. J. Friston, R. S. Frackowiak, and C. J. Price. 2004. "Structural covariance in the human cortex." *Journal of Neuroscience* 25 (36): 8303–8310.

32. C. Gaser. and G. Schlaug. 2003. "Brain structures differ between musicians and non-musicians." *Journal of Neuroscience* 23 (27): 9240–9245.

33. J. H. Kaas, M. M. Merzenich, and H. P. Killackey. 1983. "The reorganization of somatosensory cortex following peripheral nerve damage in adult and developing mammals." *Annual Review of Neuroscience* 6: 325–356.

34. D. B. Polley, M. A. Heiser, D. T. Blake, C. E. Schreiner, and M. M. Merzenich. 2004. "Associative learning shapes the neural code for stimulus magnitude in primary auditory cortex." *Proceedings of the National Academy of Sciences 101* (46): 16351–16356.

35. B. Wicker, C. Keysers, J. Plailly, J.-P. Royet, V. Gallese, and G. Rizzolatti. 2003. "Both of us disgusted in my insula: The common neural basis of seeing and feeling disgust." *Neuron 40* (3): 655–664.

36. C. Keysers, B. Wicker, V. Gazzola, J.-L. Anton, L. Fogassi, and V. Gallese. 2004. "A touching sight: SII/PV activation during the observation and experience of touch. *Neuron 42 (2)*: 335–346.

37. V. S. Ramachandran. 2000. "Mirror neurons and imitation learning as the driving force behind 'the great leap forward' in human evolution." *Edge 69*.

38. E. Altschuler, J. Pineda, and V. S. Ramachandran. 2000. Abstracts of the Annual Meeting of the Society for Neuroscience.

39. R. I. M. Dunbar and S. Shultz. 2007. "Evolution in the social brain." *Science 317*: 1344–1347.

40. B. S. Cushing and C. S. Carter. 1999. "Prior exposure to oxytocin mimics the effects of social contact and facilitates sexual behavior in females." *Journal of Neuroendocrinology 11* (10): 765–769.

41. J. L. Goodson and A. H. Bass. 2001. "Social behavior functions and related anatomical characteristics of vasotocin/vasopressin systems in vertebrates." *Brain Research Reviews* 35: 246–265.

42. L. J. Young, R. Nilsen, K. G. Waymire, G. R. MacGregor, and T. R. Insel. 1999. "Increased affiliative response to vasopressin in mice expressing the V1a receptor from a monogamous vole." *Nature 400*: 766–768.

43. L. Young and Z. Wang. 2004. "The neurobiology of pair bonding." *Nature Neuroscience* 7: 1048–1054.

44. Z. R. Donaldson and L. J. Young. 2008. "Oxytocin, vasopressin, and the neurogenetics of sociality." *Science* 322: 900–904.

45. K. D. Broad, J. P. Curley, and E. B. Kaverne. 2006. "Infant bonding and the evolution of mammalian social relationships." *Philosophical Transactions Royal Society of London Series B 361*: 2199–2214.

46. H. Walum, L. Westberg, S. Henningsson, J. M. Neiderhiser, D. Reiss, W. Igl, J. M. Ganiban, E. L. Spotts, N. L. Pedersen, E. Eriksson, and P. Lichtenstein. 2008. "Genetic variation in the vasopressin receptor 1a gene (AVPR1A) associates with pair-bonding behavior in humans." *Proceedings of the. National Academy of Sciences 105*: 14153–14156.

47. M. Kosfeld, M. Heinrichs, P. J. Zak, U. Fishbacher, and E. Fehr. 2005. "Oxytocin increases trust in humans." *Nature* 435: 673–676.

48. S. R. Donaldson and L. J. Young. 2008. "Oxytocin, vasopressin, and the neurogenetics of sociality." *Science* 322: 900–904.

49. A. E. Gandolfi, A. S. Gandolfi, and D. P. Barash. 2002. *Economics as an Evolutionary Science*. Transaction Publishers: New Brunswick, New Jersey.

50. R. V. Alatalo, A. Carlson, A. Lundberg, and S. Ulfstrand. 1981. "The conflict between male polygamy and female monogamy: The case of the pied flycatcher. *American Naturalist 117*: 738–753.

51. S. Chess and A. Thomas. 1995. *Temperament in Clinical Practice*. Guilford Press: New York.

52. See note 9.

53. K. K. Ryan and J. Altmann. 2001. "Selection for male choice based primarily on mate compatibility in the old-field mouse, *Peromyscus polionotus rhoadsi*." *Behavioral Ecology and Sociobiology* 50: 436–440.

54. H. Fisher. 2004. *Why We Love*. Henry Holt: New York.

Index

Index

Index

Index

Index